Excel by Example

Excel by Example
A Microsoft® Excel Cookbook for Electronics Engineers

By Aubrey Kagan

AMSTERDAM • BOSTON • HEIDELBERG • LONDON
NEW YORK • OXFORD • PARIS • SAN DIEGO
SAN FRANCISCO • SINGAPORE • SYDNEY • TOKYO

Newnes is an imprint of Elsevier

Newnes is an imprint of Elsevier
200 Wheeler Road, Burlington, MA 01803, USA
Linacre House, Jordan Hill, Oxford OX2 8DP, UK

 Recognizing the importance of preserving what has been written, Elsevier prints its books on acid-free paper whenever possible.

Library of Congress Cataloging-in-Publication Data

(Application submitted.)

British Library Cataloguing-in-Publication Data
A catalogue record for this book is available from the British Library.

ISBN: 0-7506-7756-2

For information on all Newnes publications
visit our website at www.newnespress.com

04 05 06 07 08 09 10 9 8 7 6 5 4 3 2 1

Printed in the United States of America.

In memory of Jonathan Moshe Kagan

Contents

Acknowledgments

The idea of this book was introduced by Carol Lewis, and her guidance and expertise have piloted it through to publication. Conversion of my manuscript to the product you have in your hands was done by Kelly Johnson. My thanks goes to them both and Tiffany Gasbarrini, and to those whose work at Elsevier has remained unseen to me, for what I hope you will agree is an outstanding effort.

I would also like to thank the management and my co-workers at Emphatec Inc. (previously Weidmuller Canada Ltd.), especially Ernesto Gradin and Don Robinson for their support, advice and encouragement for my original articles and subsequently this book.

Thanks are also due to:

Alberto Ricci Bitti for permission to use his idea, which forms the basis of Example 6, Fred Bulback for permission to include IO.DLL on the CD-ROM, *Circuit Cellar* and *EDN* for providing the format to allow me to develop my ideas and hone my writing skills.

To my children, parents and sister, all of whom encouraged me to tackle this project and whose continued interest continued to motivate, thank you.

In her usual self-deprecating manner, my wife, Nicky, has asked that she not be mentioned, and that acknowledgment is not needed for her support, both spiritual and logistical. Far be it from me to contradict her, but nevertheless, Thank You.

Introduction

When faced with a new software tool, most of us learn what we need to address our immediate problem, and then armed with 10% of the tools that are available we attempt to solve all future problems. In my discussions with colleagues, I have found that the spreadsheet is the quintessence of this effect. Almost everybody has Microsoft® Excel on their computer, yet few use it for anything but the most mundane tasks, rather like a sophisticated, but unwieldy calculator. In fact, I recently saw a newspaper article that heralded the demise of the calculator as a result of the spreadsheet, PDAs and other electronic tools.

Most of the literature on the subject of spreadsheets in general, and Microsoft Excel in particular, deal with generic cases of home economics or financial projects. Very few have direct analogies to the work done in electronics. Yet, the spreadsheet is ideally suited to allow the electronics engineer (indeed any engineer) to "work smarter, not harder." Over the years I have worked with Supercalc, Multimate, Lotus 1-2-3, Framework, Symphony, Quattro and Quattro Pro. In the end, they all are very similar. Most of what is covered in this book can be implemented in any one of the current competitors to Excel, without too many changes.

The genesis of the book was a little circuitous. My supervisor at work suggested that we should run seminars on different subjects sharing each individual's expertise. I thought some reference notes on Excel might be helpful. This led to a series of three articles that were published in *Circuit Cellar Online* starting in January 2002. Several readers contacted me and suggested additional subject matter that would be interesting. Then, out of the blue, I was approached by Elsevier to write a book based on these articles. Since the format of a book allows for more scope, I have expanded on the original ideas, added a few, and I have also tried to incorporate much of the feedback that I received.

If you only buy one book on Excel, then of course, I hope it is mine. However, it is not my intention that this book be the only book on the subject that you will ever need. I have only tried to explain general subjects that I use in the examples, since I have found them useful. I leave the detailed explanations to the more general books that are available, since I am sure they are better at it than I. Since I am writing this book for electronics engineers, I presume a degree of familiarity with a computer, including programming, and I jump into macros fairly early. I have tried to make most of the macros into a "black box" so that if you don't really want to know what goes on inside, but still need the function, you can. In addition, I have tried to make the examples "stand alone," which means that some of the basic techniques like invoking the Visual Basic® Editor (VBE) are described quite frequently.

The examples have been developed for this book under Excel 2002. No doubt by the time the book is published there will be at least one new revision. Some of the original development work was created under Excel 97, so most of this should work on any version from that time. Where I am aware that a feature has been added since '97 (such as speech input) I hope to point them out. Please forgive me if I am less than accurate with this information.

Like most of us, after a period of use I have become settled within my knowledge of the subject. I am guilty of not extending my knowledge using more of the features of Excel. Feel free to contact me and let me know what you find useful and what you think is missing. Better yet, why don't you submit the idea to *EDN* or *Electronic Design* and see your name in print (plus make a little money on the side). That's how I started; perhaps you too can write a book.

An English engineer once told me that my writing style reminded him of Somerset Maugham, a British novelist from the 1930s. This is no small feat considering that I was writing specifications for a robotic arm on the International Space Station at the time. Whilst I am sure my editor will correct all my anglicized spellings, the style will likely remain. I hope you don't find it too distracting.

It has been my experience that in any technical presentation, when the application has some glamour about it the audience is far more interested, irrespective of how mundane the technology might be. In that light, I hope that you find the ideas included in this book original, provocative and useful. Depending on work commitments, I cannot promise a speedy or detailed response, but feel free to contact me at antediluvian@sympatico.ca with comments and suggestions.

Rules of Engagement

Conventions:
I have adopted a fairly traditional approach to documenting data entry into Excel. Unless otherwise indicated, a click on the mouse is a click on the left mouse button. Notwithstanding that it is possible to change the allocation of the mouse keys, I am referring to the default configuration. Where a click of the (left) mouse button executes the desired action it is printed in bold text, for instance: **Save.** Where there is a sequence of menus that require several mouse clicks the actions are in bold and are combined by a vertical bar, for example **File | Save as**. Sometimes, a series of selections will result in the presentation of file tabs. I feel I am being consistent in documenting this click in bold as well. Things get a little greyer

when trying to describe clicking on a check box or an option button. I have tried to maintain the bold text to describe this action. Things become even murkier though when trying to describe clicking on a control that has been set up by the user. If the user has created a Combo Box and my description is to click on one of the options in the drop-down menu that appears, is it part of the application and therefore definitely in bold are part of the application and perhaps some other formatting is needed. In this set of circumstances, I don't use formatting.

Certain actions are initiated by a combination of keystrokes. These are indicated in bold with angle brackets as in **<Ctrl>.** Where there is a combination of keystrokes that must occur simultaneously, they are joined by a plus symbol. The key combination to bring up the VBA editor copy would be **<Alt> + <F11>** as an example. In Excel (and any Microsoft Office application), it is possible to run a macro from a key combination. Although this is not part of the application, this combination will appear in exactly the same way.

Any text that is entered either as data, formula or as code in VBA appears in *italics*.

VBA Help/Add-Ins:
When Excel/Office is installed, the VBA help is normally omitted. Typically, you would change this by going to Control Panel and Add/Remove Programs. Then select the Office entry. You will probably be given an option to Change the installation. Under Add features, search for the VBA help installation. On my machine, it was under the Office Shared Features folder. Select the **Run from my computer** option, and follow the prompts to install.

While you are here, also go to the **Microsoft Excel for Windows** folder in the **Add-ins** sub-folder, and set the following options to **Run from my computer** as well: Analysis Toolpak, Solver. Go one level back up the tree and enable **Text to speech** as well. Continue with the installation supplying the CDs as requested. If you don't do this, the first time you try to access one of the functions you will be prompted for the CD to complete the installation.

Analysis ToolPak Add-In:
Many of the functions that I will use in the book are available in the Analysis ToolPak add-in. You may as well go ahead and add it now or you will start to pick up #NAME errors that indicate the function was not found. This is how to do it:

1. In Excel, on the **Tools** menu, click **Add-Ins**.

2. In the **Add-Ins available** list, select the **Analysis ToolPak** box, the **Analysis ToolPak – VBA**, and the **Solver Add-in**, and then click **OK**.

3. If necessary, follow the instructions in the setup program.

Macro Protection Message:
When you first start Excel and you open a file with a macro or procedure in it, Excel will ask if you want to go ahead and do this. This is as a result of a proliferation of viruses that were passed in macros. You can modify the level of security to bypass this in Excel by following the sequence **Tools | Options | Security | Macro Security**. Choose the level (and degree of intrusion) that you are comfortable with.

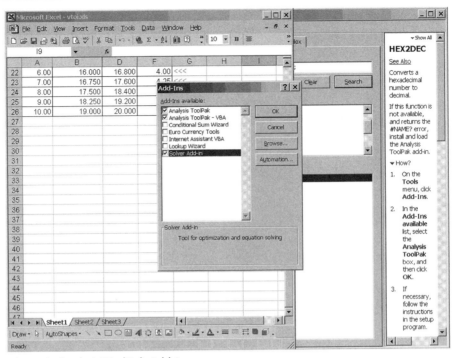

Figure 1: Analysis ToolPak Add-Ins.

VBA Variable Declaration and Naming Conventions:

They tell me that good programming practice requires that every variable be explicitly declared (in VBA using the Dim statement). VBA does not require this in its default state, probably as a hangover from the original Basic. This option can be set in the VBA environment under **Tools** | **Options** | **Editor** and select the **Require Variable Declaration** option.

It is also convention to follow Hungarian notation when naming variables and objects. In this method, object names are prefixed by a three-character identifier (Form1 would be called frmForm1), and a variable would be prefixed by a single character that identifies its data type. iVariable would be an integer.

Some of my examples were developed before I was aware of this notation and in others I simply forgot or was not disciplined enough to employ it. In addition, most of my programming is self-taught and was based on small microcomputers, so I am sure I commit all manner of software coding sins from public variables to goto statements to insufficient comments. I am afraid it is very much a case of "Do as I say, not as I do."

Documenting Worksheets:

I find documenting Excel quite difficult. Normally, you only see the end result while there is actually a formula behind the result and there is a "knock on effect" as results depend on other cells. Formatting is even more difficult because it may not be obvious that the cell has been formatted. There are techniques in Excel to unmask these hidden factors, but they require explicit actions. Use the **Formulas** option in **Tools | Options | View** to see the formulas used. Antecedents and precedents can be traced using the **Tools | Formula Auditing** sequence. Conditional formatting can be identified by clicking on any cell and then **Edit | Go To | Special | Conditional Formats**. It is possible to find conditional formatting like the current cell or all conditional formatting. Pay attention to the other **Go To** options here. All comments can be made visible with **View | Comments**. To find out what range a name refers to, use the drop-down arrow by the name box to find the name and then click on the name.

What's on the CD-ROM?

Included on the accompanying CD-ROM:

- A full searchable eBook version of the text in Adobe pdf format
- Ready to run, customizable Excel worksheets for each application covered

EXAMPLE 1

Voltage-to-Current Converter

Model Description

A very common use of Excel is to enter data into the worksheet and then use its power to analyze the data. To start with, let me present a simple application to demonstrate some of the tools included with Excel to enhance productivity.

In the industrial automation field, analog signals are still distributed using the venerable 4 to 20 mA current loop. In this technique, the output from any transducer is conditioned by means of some electronics to generate a current of 4 mA at the bottom of the scale and vary continuously up to 20 mA at the top of the scale. The block diagram in Figure 1-1 describes just such an application where a 0 Vdc to +10 Vdc input signal is translated to 4–20 mA output. The transfer function is $I_{out} = ((V_{in}/V_{fullscale}) * 16) + 4$. The current, I_{out}, is measured in milliamps. Since $V_{fullscale}$ is 10V, this function reduces to $I_{out} = 16 * (V_{in} /10) + 4$.

Figure 1-1:
0 Vdc to +10 Vdc is converted
to 4–20 mA through this module.

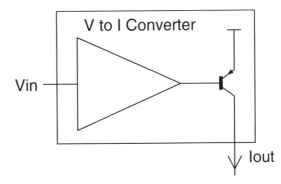

This application will take the measured input voltages and output currents and analyze the linearity of the system. Even though this is a simple application, I use it quite frequently. It is useful to build a model since the measurements are taken at different ambient temperatures to establish performance specifications. In this example, the data is keyed in by hand. It is simple enough to do, but we will see in a later application how the data can be acquired automatically.

Starting Excel

This early example includes several very basic features in Excel that I normally find useful. In many cases they are intuitive, in others they may be well known to most of us. Nevertheless, I think it is beneficial to go through them, slow though it may be. Please forgive me. We will move a lot faster in later examples.

When we first open Excel 2002, there may be an extra window on the right side of the screen that simplifies the creation of a new file as shown in Figure 1-2. I am not partial to this screen, so this is the only time you will see it in this book. We can get rid of it by unchecking the appropriate box, or we can turn it on or off through the menus: **Tools | Options**, and in the **View** tab, select or deselect the box named *Startup Task Pane*.

As with most applications, we can close the window using the **X** in the top right-hand corner.

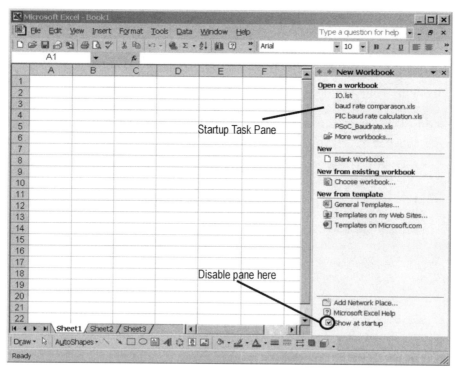

Figure 1-2: Startup screen in Excel with Startup Task Pane.

Excel refers to a spreadsheet as a *workbook*. Inside the workbook there can be several sheets (referred to as *worksheets*), and there can be several workbooks open at a time. Open a new workbook. If the Startup Task Pane is still available, simply click on the "Blank Workbook" selection. Otherwise, we can start a new workbook in several ways depending on our preference. There is the menu option: **Files | New** selection, the keyboard shortcut (**<Ctrl> + <N>**) as listed in the menu option, and there is also an icon on the extreme left of the toolbar (see Figure 1-3).

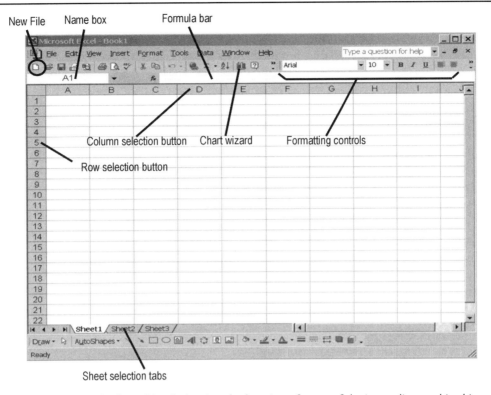

New File   Name box   Formula bar

Column selection button   Chart wizard   Formatting controls

Row selection button

Sheet selection tabs

Figure 1-3: Blank workbook showing the location of some of the items discussed in this example.

Data Entry into a Worksheet

Entering information into any cell of the workbook is easy. Excel automatically presumes any data is text and left justifies it. Any information that is purely numerical is interpreted as a number and is entered right justified. Any entry preceded by a mathematical symbol "+", "–" or "=" is interpreted as a formula. A number can be manipulated directly using formulas and so forth, whereas text is normally processed through string manipulation in formulas. Of course, it is possible to change the justification as well as the format of a cell or a group of cells using formatting controls and the standard Microsoft® Windows® techniques. If we want an entry to be interpreted as a string when the default will interpret it as a number, we prefix the entry with an apostrophe ' . If we enter a number that includes nonnumeric characters, Excel will simply identify it as an error.

If a text entry in a cell is too long to fit within the cell it will appear to flow over to the adjacent cell, if that cell is empty. If the adjacent cell is not empty, the text will be truncated. There are some techniques that allow us to improve on this which we will see later on. Editing the entry though still requires clicking on the original cell.

It is possible to size the width and height of columns and rows by moving the cursor over the line that demarcates the separation of the columns in the column selection bar (or rows in the row selection bar) until the symbol changes to a bar with arrows on either side and then clicking and dragging the line. One of the secrets of Windows is that it is possible to automatically size the column (this works in Windows® Explorer as well). Instead of clicking and dragging the line, simply double-click when the double arrow symbol appears. The column will adjust to fit the largest entry in that column.

Entering data into a cell is intuitive, but there are some features that can simplify the process. It is possible to terminate the entry by pressing the **<Enter>** key or one of the direction arrows. The arrows are a kind of shorthand in data entry, so that one keystroke both enters the information and takes us to the next cell (in the direction of the arrow). Actually, they only work on the original entry of data and not when the data is edited. The *Tab* key will also enter the data and move one column to the right. The action of the **<Enter>** key after using the **<Tab>** is quite interesting. In this case, Excel determines that we are entering tabular data and the **<Enter>** key will vector us to the cell below the cell we started the horizontal data entry. This is great when we are entering several columns of data, line by line as it acts like a carriage return and line feed.

Aside from the Tab technique above, it is possible to decide which way the cursor will move after *Entering* information. It can be changed in the **Tools | Options** menu and the **Edit** tab.

Editing data in a cell requires that we click in the cell and then click in the formula bar in order to edit there. Alternatively, we can double-click on the cell in question, and edit directly in the cell. We can now resort to the usual editing procedures.

Let us view some of this in action. Open a new workbook. In cell A1, enter the text *Voltage to Current Converter*. Now navigate back to cell A1 using the direction arrows or more simply, click on cell A1. It is possible to format the appearance using the controls on the task bar or the menu controls. Change the font size to 12, and make the appearance bold using the drop-down box by using the format option on the task bar.

Click on cell A5. Enter the text *Input Voltage* and press the Tab key. Notice how it overlaps into B5. In cell B5 (which should be selected already), enter the text *Output Current* followed by the Tab key. The text in cell A5 is now truncated. Enter the text *Proportion* in cell C5 followed by the Tab key. Enter the text *Theoretical Current* in cell D5, followed by the Tab key. In cell E5, enter the text *Error* followed by the Tab key. Finally, enter *Linearity Error*, followed by the Enter key in cell F5. Note that the selected cell is now A6. The screen should look like Figure 1-4.

Right click on the row select button for row 5. Click on **Format Cells** and then click on the **Alignment** tab. In the **Text Control** section, check the **Wrap text** option and then the **OK** button. Move the cursor so that it hovers over the line demarcating two columns in the column selection bar and drag the line so that the text splits into two columns with the desired visual effect.

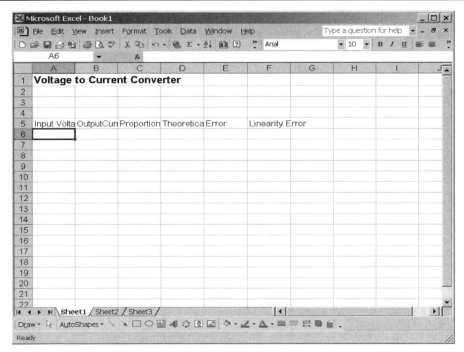

Figure 1-4: Table headings showing the truncation of the text.

Move the cursor so that it hovers over the join of columns A and B in the column selection bar. Double-click and notice how the column width changes. It actually changed to accommodate the widest entry in the column, the title in cell A1. Return to the join of columns A and B and click and drag the line so that column A is a suitable width. Of course, the Undo feature could also be used.

Let's do a little more bulk formatting. Click on the row selection button for row 5 on the extreme left of the screen. Notice how the whole row is selected. Now click on the **B** (for Bold) button on the task bar and the whole row is instantly converted to a bold type.

Autofill

Let's assume that we are going to apply 0 volts to 10 volts at the input of our conversion module in steps of 1V. This data will be entered in column A. Now we could simply type in 0 in A6, 1 in A7 and so on all the way through to 10 in cell A16. This could prove tedious, and Excel provides an extremely easy method of autofilling. Enter the value 0 in cell A6 and 1 in cell A7, providing the seed of the starting number and the increment (or decrement) for the autofill. Now click on A6 without releasing the mouse button and drag to cell A7 so that both are selected as shown in Figure 1-5.

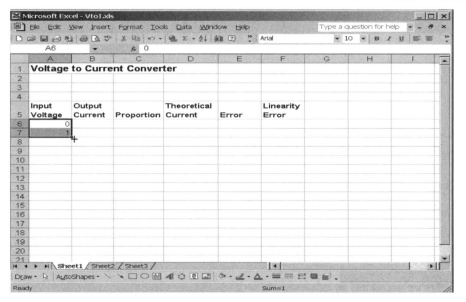

Figure 1-5: Preparing for autofill. Cell A6 is the initial value, and the difference between A7 and A6 provides the step increment.

Notice the little black square (called the *fill handle*) at the bottom right-hand corner of the frame surrounding the selection. Move the mouse cursor over this square and the cursor will transform from a large unfilled plus sign to a smaller solid plus sign. When this happens, click and drag the fill handle down. Notice a small yellow box that pops up near the cursor. This is the maximum number that will be "autofilled" when we release the mouse button. Drag this until the pop-up number reads "10" as in Figure 1-6 and release the mouse button.

In Parenthesis: *Copying With and Without Format*

There is a small symbol at the bottom corner of the block that has just been filled. This symbol is only present on later versions of Excel. Clicking on it allows us to modify the way the formatting of the cells involved in the autofill is affected. We can safely ignore it for the moment.

In earlier versions of Excel (and Microsoft® Word), there was an option of copying with or without the formatting, but it was buried in the menu system. This is now more explicit as can be seen by this and similar symbols that appear on the worksheet after an autofill or copy operation. It can, however, be disabled in the **Tools | Options** menu and the **Edit** tab.

While we build the model we don't have any real data, but it helps to have some numbers to work with. Enter 4 representing the intial current of 4 mA in cell B6 and 4.5 in B7. Autofill using these cells to cell B16.

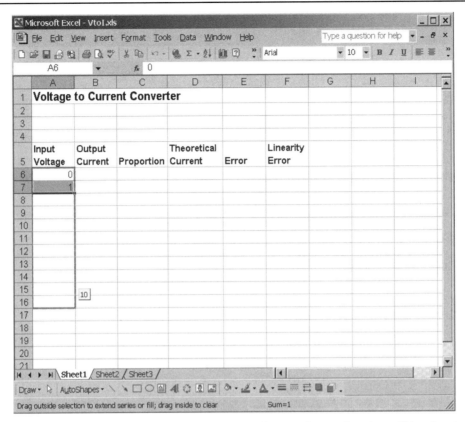

Figure 1-6: Autofill in action. Note the pop-up showing the number that will be placed in the last cell of the selection. Releasing the button at this point will result in the entries running from 0 to 10 in unit increments.

In Parenthesis: *Autofill of Nonnumeric Sequences*

The autofill function is quite intelligent and will recognize dates, days and some other normal sequences. If you enter a series of text entries, the autofill will fill the range with the same text in the same order.

Bulk Formatting

The formatting of the cells in column B is not fixed, so the number of digits changes and the appearance is unappealing as well as philosophically incorrect since the digital ammeter that we will be using reports the current to 3 decimal places. We can bulk format this as well. Click on the "B" column selection button and the whole column is selected. Now right click on it to bring up some options (actually we could achieve the same thing by simply right-clicking on the "B" column selection button). Select the **Format Cells** option and then the **Number** tab. In the scroll-down menu, click on **Number** and set it to 3 decimal places and click on **OK**. I know that, depending on the accuracy of the meter, we shouldn't place too much trust in the last digit, but that is another issue. Even though the column heading is in text, Excel is smart enough to leave its formatting alone, modifying only the numerical entries.

If a column is too narrow to display a number in the selected format, Excel resorts to displaying the symbol "#####." To see this effect, move the join of columns B and C so that B becomes too narrow to display the full number. Resize the column by double-clicking the column join.

We will be measuring the input voltage on a digital voltmeter with 3 decimal places as well, so format column A to 3 decimal places as before.

In Parenthesis: *Multiple Selections*

It is possible to select many groups of cells for bulk actions using standard Windows techniques by using the *Ctrl* and *Shift* keys in conjunction with the mouse block selection.

Formulas

In order to calculate the theoretical current output for a given input voltage, I am going to break the calculation in stages. As with all programming, it is possible to embed calculations within calculations and it quickly confuses things. Maintaining embedded calculations (although providing job security) can be troublesome. For instance, try figuring out what

$=IF((DCOUNT(AL7:AX31,1,AX33:AX34)>0),DAVERAGE(AL7:AX31,1,AX33:AX34),"")$

refers to, especially four years after you have written it.

The model we are developing is a simple example, and several of the steps could easily be compressed, but this will allow me the opportunity to demonstrate some techniques to improve readability and reliability.

The third column was entitled "Proportion," and this calculation will result in the proportion of the measured input to the full scale. In other words, an input of 3 volts on a span of 10 volts (0 to 10) would give a number of 3/10. So the calculation would be:

$$(V_{in} - V_{min})/(V_{max} - V_{min})$$

where V_{min} corresponds to the minimum input voltage of 0V, and V_{max} to the maximum input voltage, 10V.

To enter this in the worksheet, click on cell C6. Now enter the following:

$=A6/A26$

The "=" sign must be there (or a "+") to indicate that it is a formula. A6 is the input voltage in this column, and A16 contains V_{max}. Notice how this appears in the formula bar at the top of the worksheet and in the cell, until we press **<Enter>** (or of course, the direction arrows or the Tab key). The cell shows the result of the calculation. When we click on the cell, the formula will appear again in the formula bar. Figure 1-7 shows where we're at.

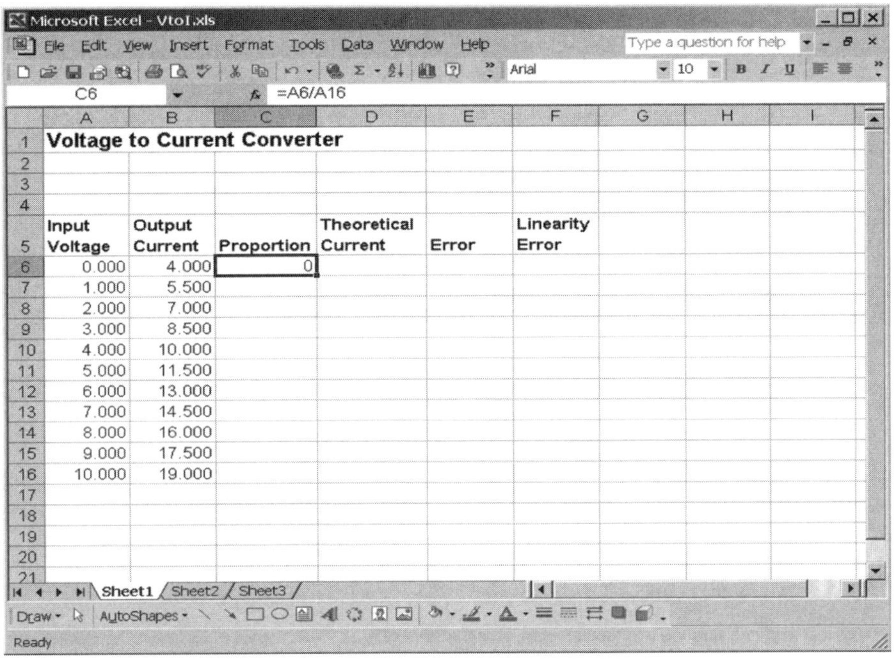

Figure 1-7: Result of formula calculation in cell C6, and the formula used in the calculation appears in the formula bar.

Copying Formulas

We could now enter this formula for every cell in the column, but of course it is simple enough to copy and paste. Click on cell C6 and copy to the clipboard with standard Windows techniques such as **<Ctrl> + <C>.** Now click on cell C7 and paste. Ignore the error that appears for the moment and click on C7 to check the formula in the formula bar:

=A7/A17

Excel has made the relative translation for the cell and this is very handy *except* that A17 does not contain V_{max}.

In Parenthesis: *Adding Columns/Rows*

It is an easy matter to insert a column or a row. Simply use the menu function **Insert | Column** and everything will move to the right by one column. Inserting a row is very similar. All the relative addresses and absolute addresses are all maintained without missing a beat.

> **In Parenthesis:** *Deleting Columns/Rows/Cells*
>
> The delete cells function is very similar to adding a column/row, except we go through the **Edit | Delete** menu and then decide whether we need to delete a column or a cell and what the result will be. Of course, if you delete an entity that is being referenced elsewhere, Excel will report a problem. Fortunately, there is always the Undo button.
>
> Depending on how a cell is deleted influences the residual effect on the cell. If we select the cell and press **Delete** or **Backspace**, the contents are gone, but the formatting remains. To get rid of the formatting, we use the menu action: **Edit | Clear | Formats.** To clear both, use the menu action: **Edit | Clear | All.**

Relative and Absolute References

Obviously, we need some technique to reference a particular cell without allowing for the automatic adjustment. Simply entering the cell coordinates is known as *relative referencing*. In order to make an absolute reference, use the prefix "$" to the column or the row identifier, or both. The ability to create an absolute reference in one dimension only can prove very handy.

Double-click on cell C7 to edit the formula in the cell or click on the formula bar to edit it there. Modify the formula to read:

=A7/A$16

The "$" symbol only references the lines since there is no copying across the columns yet. We could have used the following format to fix the changes in either dimension:

=A7/$A$16

Terminate the entry and then copy cell C7 to the clipboard. Block the range C8 to C16, by clicking on the former and dragging to the latter, then paste. Figure 1-8 should be the result.

Clicking on any one of the entries in the column will show that only the first cell reference is relative. The numbers look reasonable, but not formatted. I am not going to bother doing that at the moment. This is a transitional calculation and will be hidden later.

The next stage of the calculation is to take the number from each cell in the Proportion column and multiply by the range of the output, 16 mA and add the offset of 4 mA. Click on cell D6 and enter the following:

=(c6*16)+4

It should result in a calculated value of 4. Now copy this cell to the range D7 to D16. The entry in D16 should be 20 corresponding to the top end of the output current range. Format the column for 3 decimal places. Figure 1-9 is the result.

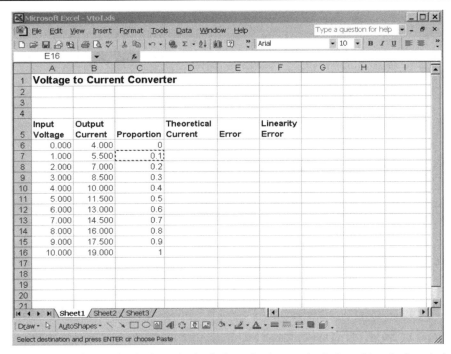

Figure 1-8: The result of a formula (including absolute and relative addressing) copied to a block of cells.

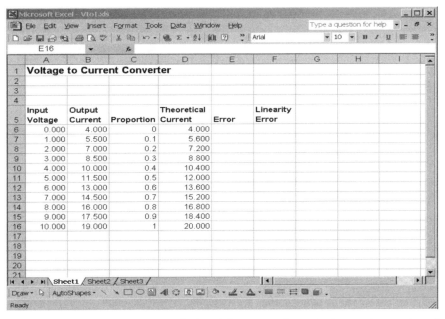

Figure 1-9: Theoretical current calculated from a mathematical operation on the cell in the previous column.

The calculation of the error of the measured output compared to the theoretical output is simple, subtracting the measured value from the theoretical value. Click on cell E6 and enter the formula:

> =d6-b6

and copy this to the other cells E7–E16.

The definition of linearity is not absolute. Generally, it is given as a percentage based on calculation of the error divided by the full scale value, although on 4–20 mA, it could be argued that the maximum range is only 16 mA. Opting for former definition, we enter:

> =E6/FullScale

in cell F6. This is also copied to the range F7 to F16.

There is an alternative method to fill a block of cells. To see a description of the feature, see "In Parenthesis: Fill Box" in Example 16.

Enter the text *"Full Scale Value"* in cell A3. In B3, enter the value "20". It will take the 3 decimal place format already existing for the column. In order to fit the text in A3 into the cell, another option is possible. Right-click on the cell, select **Format Cells** and click on the **Alignment** tab. In the **Text Control** section, check the **Shrink to Fit** option, followed by the **OK** button.

Naming Cells

Any reference to the Full Scale Value would be an absolute reference to B3, that is $B$3. On any complex worksheet, trying to skip back and forth trying to figure out what $B$3 refers to is exceedingly inefficient. Excel allows us to provide a contextual name for either a cell or a group of cells. It can be done in two ways. The first is more generic, and is helpful where there are multiple pages or multiple workbooks. It is accessible through the menu sequence:

> **Insert | Name | Define**

It is possible to navigate to the cell or range of cells within the dialog, although it does conveniently start at the current cell selection. We will need this dialog to remove a name.

I find the second method quicker for simple applications. Simply select the cell (or group of cells) on the worksheet. In the current example, this would be cell B3. Then click in the Name box at the top left-hand corner of the spreadsheet (see Figure 1-3) just to the left of the formula bar. Enter the name of the cell here, *FullScale*, with the result appearing in Figure 1-10.

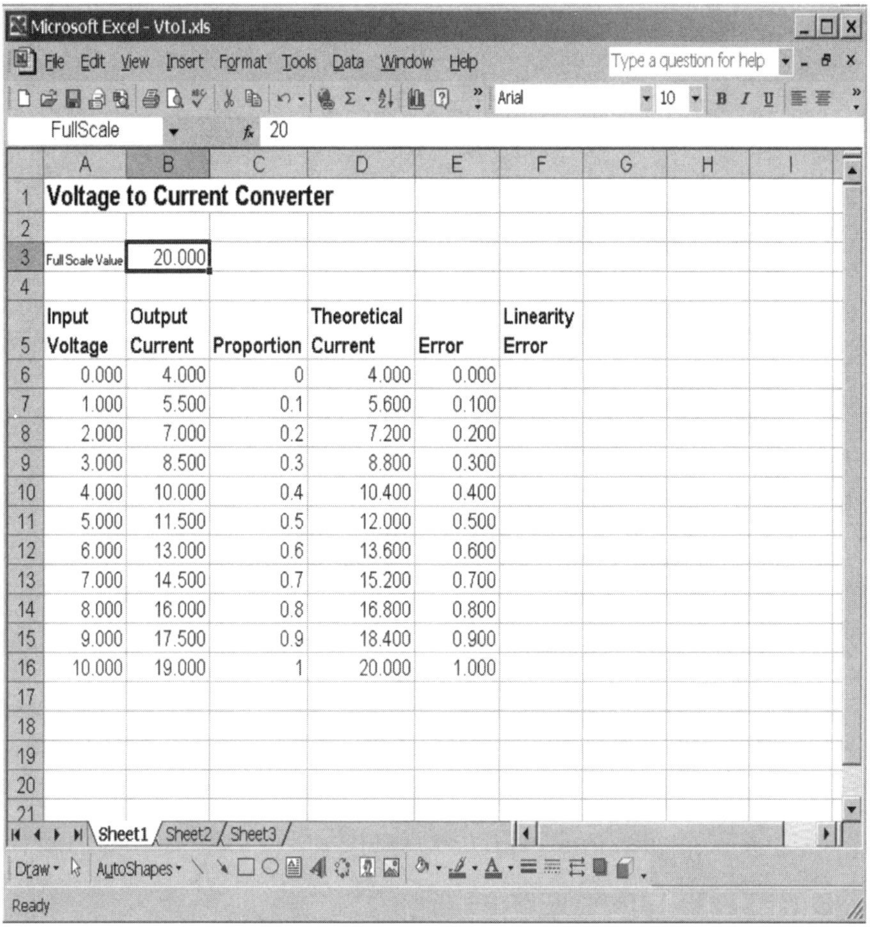

Microsoft Excel - VtoI.xls

FullScale | 20

	A	B	C	D	E	F
1	Voltage to Current Converter					
2						
3	Full Scale Value	20.000				
4						
5	Input Voltage	Output Current	Proportion	Theoretical Current	Error	Linearity Error
6	0.000	4.000	0	4.000	0.000	
7	1.000	5.500	0.1	5.600	0.100	
8	2.000	7.000	0.2	7.200	0.200	
9	3.000	8.500	0.3	8.800	0.300	
10	4.000	10.000	0.4	10.400	0.400	
11	5.000	11.500	0.5	12.000	0.500	
12	6.000	13.000	0.6	13.600	0.600	
13	7.000	14.500	0.7	15.200	0.700	
14	8.000	16.000	0.8	16.800	0.800	
15	9.000	17.500	0.9	18.400	0.900	
16	10.000	19.000	1	20.000	1.000	

Sheet1 / Sheet2 / Sheet3

Figure 1-10: Naming cell B3 to "FullScale".

In Parenthesis: *Cell Names*

You need to exercise a little caution in naming cells using the name box because the name box also serves as a GoTo box. Clicking on the box and entering a valid cell coordinate will result in the addressed cell being selected and not the current cell being renamed. You will see this later when trying to name a cell for a resistor, for example, R17. There are a huge number of possible columns in Excel. They range from A through to Z, AA on past AZ followed by BA and on until you get to column IV. On a positive note, the drop-down control at the side shows all the named cells and clicking on that will take you to your destination rather quickly. There are rules for naming cells as you would expect. The name cannot start with a number, nor can it include spaces or other special characters. Reference to the cell name is not case sensitive.

Using a cell name of course is an absolute reference.

In Parenthesis: *Worksheet Navigation*

While we are on the subject of moving around the workbook, aside from the Name box, it is possible to go to a cell location using the **Edit | Go To** menu sequence or **<Ctrl>** + **<G>** will bring up the same dialog.

Ctrl plus the direction arrows, Home or End, result in grand movements. They are sometimes contextual, depending on where you are in a worksheet. For instance, if you have cell A1 selected in the spreadsheet we are currently working in and you press **<Ctrl>** + **<→>** you will arrive at cell IV1. If you are in the body of the worksheet, cell C8 say, the same key combination will take you to the last valid column of data. Repeating the key combination will then take you to IV8.

<Ctrl> + **<Home>** returns you to cell A1.

Any reference to the cell can simply use the name as a handle. Click on cell F6. Enter the formula:

=(*e6/fullscale*)

That is the error divided by the full scale range. Now copy this cell from F7 to F16 and then format column F as a percentage to two decimal places by right-clicking on the column F selection button, choosing **Format** and selecting the **Percentage** entry under the **Number** tab. The reference to FullScale is an absolute reference. Figure 1-11 should be the result.

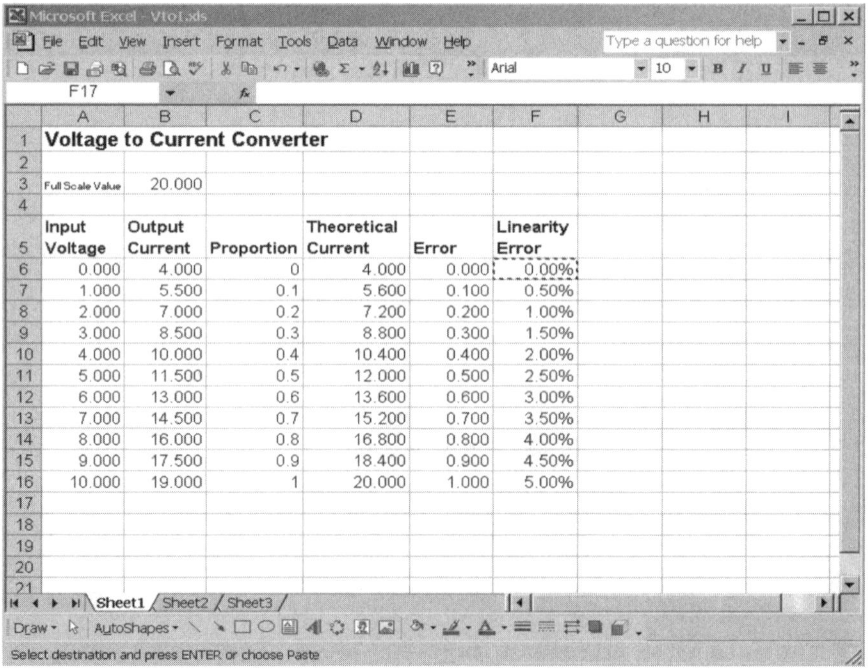

	A	B	C	D	E	F
1	Voltage to Current Converter					
2						
3	Full Scale Value	20.000				
4						
5	Input Voltage	Output Current	Proportion	Theoretical Current	Error	Linearity Error
6	0.000	4.000	0	4.000	0.000	0.00%
7	1.000	5.500	0.1	5.600	0.100	0.50%
8	2.000	7.000	0.2	7.200	0.200	1.00%
9	3.000	8.500	0.3	8.800	0.300	1.50%
10	4.000	10.000	0.4	10.400	0.400	2.00%
11	5.000	11.500	0.5	12.000	0.500	2.50%
12	6.000	13.000	0.6	13.600	0.600	3.00%
13	7.000	14.500	0.7	15.200	0.700	3.50%
14	8.000	16.000	0.8	16.800	0.800	4.00%
15	9.000	17.500	0.9	18.400	0.900	4.50%
16	10.000	19.000	1	20.000	1.000	5.00%

Figure 1-11: Full model.

Hiding Cells

As I promised earlier, columns C and E are intermediate steps and so do not need to be visible. Column select both by clicking on column select button C and then pressing the <Ctrl> key and clicking on column select button E. Then right-click on either and select the **Hide** option. See Figure 1-12.

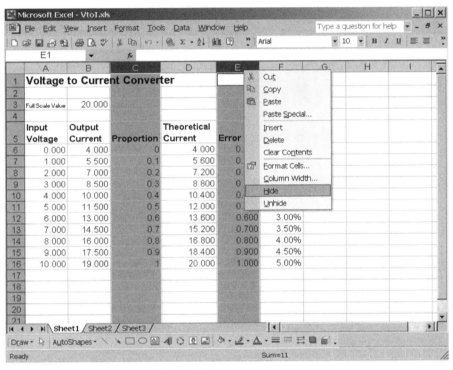

Figure 1-12: The process of hiding cells.

The columns now disappear, although their presence is denoted by the discontinuity of the column lettering. It is also indicated by the thickening of the divider line on the column buttons. To retrieve the column, select the columns on either side of the "divide." Simply click on the one column adjacent to the split and drag to the adjacent column on the other side. Right-click and select **Unhide**. Multiple selections with the <Ctrl> key will not work successfully to unhide the hidden column.

Borders

In order to enhance the tabular appearance, we must add lines around the entries. Block the area where this formatting is to be done as shown in Figure 1-13.

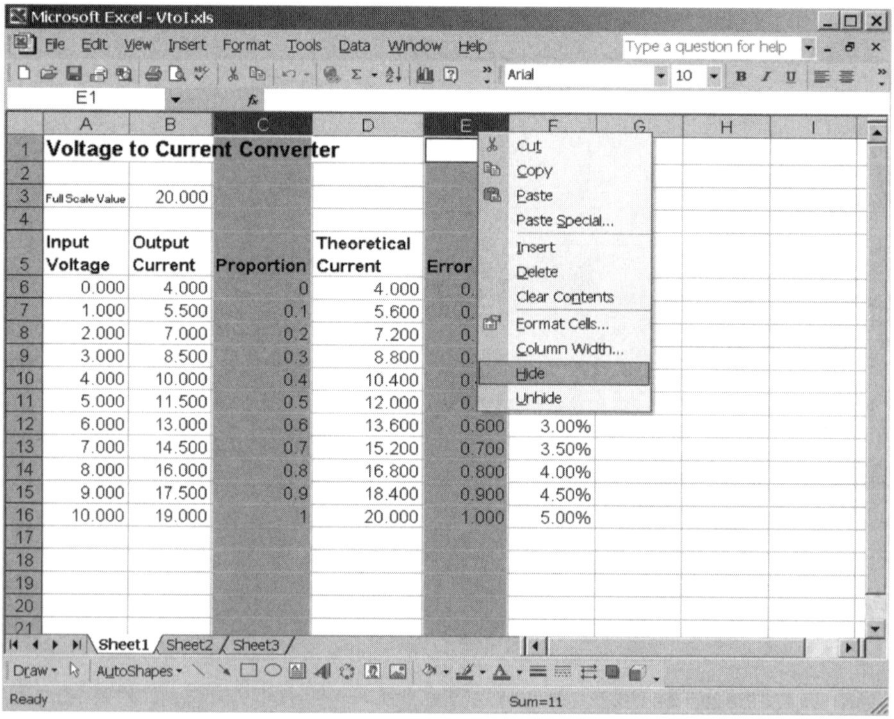

Figure 1-13: Preparing to add borders.

We can do this in a number of ways. In this simple case, it is possible to click and drag from cell A6 to F16. Where a table is bigger, this can sometimes be inconvenient, because, as you will discover, dragging the selection over a large number of columns or rows results in them whizzing past. It becomes difficult to get to exactly where we want to go.

Click in cell A6 and then navigate using the control bars till the last cell is visible. Press the *Shift* key and click in the last right-hand cell of the table. This too, can be inconvenient if the table is large since navigating to the last cell can lose the initial cell selection. Here is a really quick way. Click in cell A6. Use the key combination **<Shift> + <Ctrl> + <End>** and our selection is done.

Back to the job in hand. Within the selection, right-click and select **Format Cells** and the **Border** tab. For the current selection, the dialog allows us to select line widths, which side to have a line, hatching and cross hatching and many other options. Note that we could do this for any group of cells starting from an individual one. While we are here, notice the other tab options that allow background color to be modified, cell alignment, text fonts and so on. Return to the **Border** tab and click first on the line style—I am using the lightest weight (bottom left-hand corner). Then click on **Outline** and **Inside** and the **OK**, and the model is ready for operation.

Bells and Whistles

While we are here, why don't we add some features? Most of the time we only do what is necessary to complete a job, but I want to whet your appetite for some of the subjects that we will cover later.

Conditional IF and Absolute Value

Let's set the upper limit of linearity performance to ±1%. Instead of scanning each result, why don't we add a marker on the right of the table to indicate when the reading is out of limits?

Since the error could be positive or negative, we need to look at the absolute value. The ABS function works exactly as if we are working in a computer programming language (which is of course what we are doing) returning the absolute value of the number it is handed as a parameter. The IF statement is perhaps a little more cryptic than the IF, THEN, ELSE construction of a high-level programming statement, but that's what it does. It has three parameters separated by commas. The first parameter is the logical test, the second is the value of the cell if the test is true, the third if the value of the cell if the test is negative.

Click on cell F3 and add the text "*Maximum Tolerance.*" In cell H3, add the value *1* and name the cell *MaximumTolerance*. Format the cell for percentage. Now click on cell G6 and enter the formula:

$$=IF(ABS(F6>Maximum Tolerance),">\!<\!<\!<",">")$$

In other words, if the value in cell F6 is greater than the value in MaximumTolerance (cell H3), the symbols <<< will appear in that cell.

Copy cell G6 to cells G7 to G16. Then format cells G6 to G16 so that the text is red (by formatting the font). I deliberately chose the output current results so that there would be faults. From row 9, all the readings should be indicated by the <<< symbols in column G. The advantage of using a constant in the worksheet is that if the specifications change allowing and easing of the linearity, or tightening the requirements, it is a simple matter to go to cell H3 (using the Goto MaximumTolerance sequence, if you like) and modify the value to whatever is required.

Chart

The charting application can easily be used in any workbook. Let's get a feel for it now. Block from cell A5 to B16 and then using the **<Ctrl>** create a second block from D5 to D16. Actually, if column C is hidden, we can create as a single block, since hidden columns won't be used in a chart. Be sure to include the headings of the columns as the charting wizard will use this information to identify the curves. Click on the Chart Wizard button on the toolbar or go through the **Insert | Chart** menu. We will be presented with a dialog box as shown in Figure 1-14.

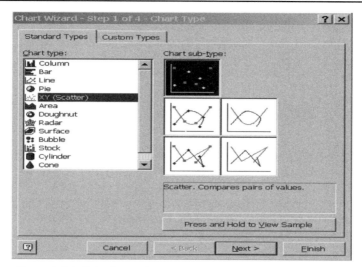

Figure 1-14: Selecting a chart type.

There are many possible types of charts. The initial charts and defaults are more for marketing types. As Dilbert might say, "So that they can find them!" As engineers, we will mostly need an XY (Scatter) chart. Select this chart type, and then click on the sub-type on the bottom left. Click on the **Next** button.

The next screen (Figure 1-15) gives an idea of what the chart will look like and an opportunity to change the selections that Excel has guessed at based on the initial selection (if there was one). Clicking on the **Series** tab allows us to select additional ranges as well. Click on the **Next** button.

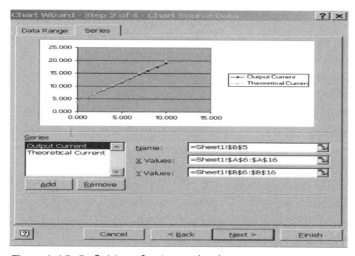

Figure 1-15: Definition of series on the chart.

In the next step, we can add some cosmetic effects, labels and gridlines amongst them. When we are satisfied with the settings shown in Figure 1-16, click on the **Next** button.

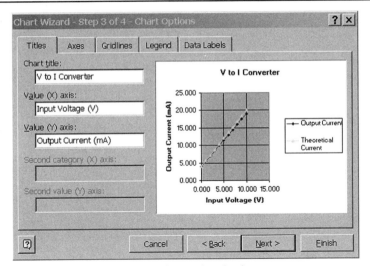

Figure 1-16: Adding chart titles, and so forth.

As the final step of the wizard, we have to decide if the chart goes in the same worksheet as the data or in a sheet all on its own. For the time being, place it in the current worksheet. We get to select exactly where and how big, by dragging it around and sizing it.

I have sized it quite large in order to magnify the separation between the two plotted lines. See Figure 1-17.

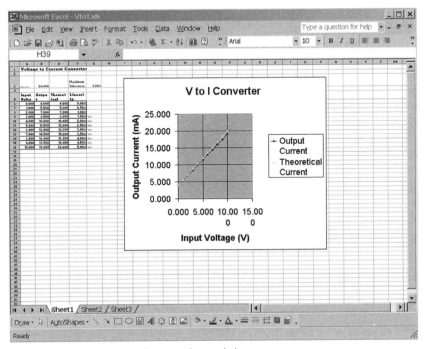

Figure 1-17: Placing the chart on the worksheet.

> **In Parenthesis:** *Zoom*
>
> You can zoom in or out using the **View** | **Zoom** menu, or pressing Ctrl and using the scroll wheel on the mouse.

By clicking on each element of the chart, it is possible to change some associated properties. For instance, the input voltage will never go above 10V, so by clicking on the horizontal axis, this can be modified. The vertical axis can be changed in a similar manner.

Error Bars

It is possible to add error bars on a curve in a chart. Click on the Output Current line in the chart and in its **Format** option deselect the markers (under the **Patterns** tab). Return to the worksheet by entering the heading "*Error*" in cell E5. In cell E6, enter the formula:

   =*MaximumTolerance*FullScale*

We use this calculation to find the maximum allowable error. Now copy this formula from cell E7 to E16. It is the same value, but for what I am going to show you, Excel needs the information in this form.

Return to the chart and click on the Theoretical Current line in the chart, deselect the markers and then select the **Y Error Bars** tab as in Figure 1-18.

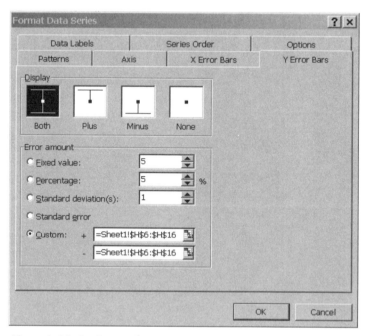

Figure 1-18: Setting up the Y Error Bars.

Click on the **Custom** option, and using the block select button on the right of the data entry box we can visually block the data for both the positive and negative sides of the error bar.

Now whenever the measured value is outside the theoretical value, it is visible on the chart. The error bars are visible in Figure 1-22.

Adding a Trendline

Excel has the capability of generating a regression on a line and determining the best form of the equation. Right-click on the Output Current line and select **Add Trendline** from the menu as shown in Figure 1-19.

Figure 1-19: Initiating a trendline.

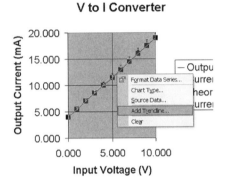

Figure 1-20 shows the dialog that appears. Select a linear regression and then click on the **Options** tab.

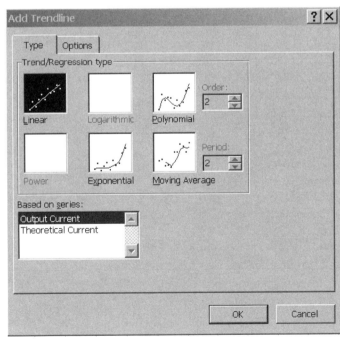

Figure 1-20:
Defining the type of regression.

Under the options (Figure 1-21), ensure the **Display equation on chart** is selected and click on **OK**.

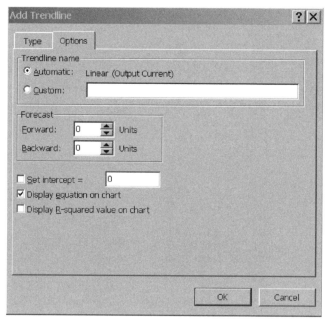

Figure 1-21: Adding the equation to the chart.

The trendline and the equation for it are added to the chart as shown in Figure 1-22.

Figure 1-22:
Trendline and associated equation.

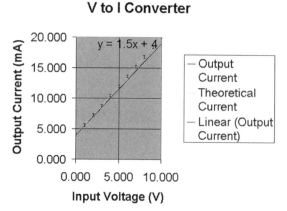

Macro: Timer

In many circuits, especially where a current is being generated, the output requires some time to stabilize. Why not use the computer to time this period and notify us when it is time for the next measurement?

In order to do this, we need to use a macro. I will go into this in greater depth later in the book. For the moment, just follow along to see the effect. In order to use this feature, we will need to have installed certain add-in modules as described in the front of this book.

Navigate the menus **Tools** | **Macros** | **Visual Basic Editor**. We should arrive at a screen resembling Figure 1-23.

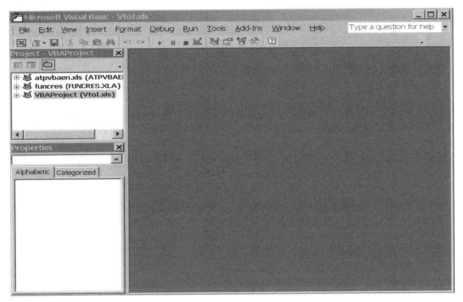

Figure 1-23: Startup Visual Basic screen.

Click on the workbook name in the Object Explorer window (top left). If it is not visible enable it through the menus **View** | **Project Explorer**. Now click on **Insert** | **Module** and something like Figure 1-24 should result.

At the bottom left of the module window, click on the **Full Module View** button, that is, the one next to the arrow. We should see a tool-tip pop up with the information.

Click in the module window and type:

sub Notify

followed by the **<Enter>** key. This immediately notifies Visual Basic to create a template for a new procedure (also known as a subroutine in older parlance). Visual Basic will automatically add some formatting and the last line "End Sub". Click after the End Sub line and type:

sub StartTimer

followed by the **<Enter>** key.

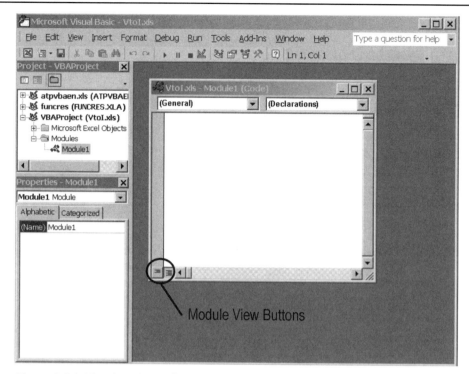

Figure 1-24: Visual Basic, ready to enter code.

Notice how Visual Basic prepares the second template and delineates it from the first. Click between the module view buttons to see the effect. Obviously, the Procedure view shows only one procedure at a time.

Now add the text for both procedures as in Figure 1-25. The text in Notify produces a message box to notify the user that the time has elapsed. It is accompanied by the OK button with this message box, plus a sound tone to attract the user's attention. As we enter the text, we notice the Visual Basic prompts guiding us for the next parameter or format required.

The StartTimer procedure initiates an action to take place 1 minute after the procedure (which will be our macro) is run. That action is to run the Notify procedure. It will continue to run even if we move to another application in Windows while we wait. It appears to me that the version of Windows determines the exact action that occurs when in another application and the timer has expired, but we will be notified somehow. In Windows 2000, the Excel button in the desktop task bar begins flashing.

Return to Excel by clicking on the Excel button on the left of the toolbar, or by any other technique including the Windows taskbar. Once in Excel, click on **Tools** | **Macro** | **Macros** to arrive at Figure 1-26. Notice both procedures appear as macros. Click on StartTimer. It is possible to click the **Run** button now to run the macro, but to do this every time will become tedious.

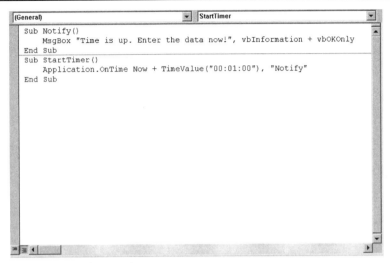

Figure 1-25: Timer code.

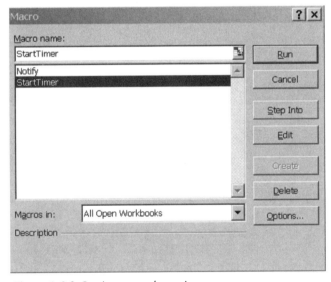

Figure 1-26: Setting up and running a macro.

There are many techniques to invoke macros, but for the moment let's use a shortcut key. Excel itself uses a number of shortcut keys (**<Ctrl> + <X>**, for example), so we may not want to use some of these. It is also possible to create a **<Ctrl> + <Shift>** combination. Click on the options button, and in the next window press the **<Shift> + <G>.** Click on the **OK** button to return to the window in Figure 1-26. Close this window.

And that does it. To take a reading, hit the shortcut combination **<Ctrl> + <Shift> + <G>** and wait to be prompted to take the next reading.

The file named *VtoI.xls* can be found on the CD-ROM accompanying this book.

EXAMPLE 2

Baud Rate Selection

Model Description

Most 8-bit microcontrollers on the market today include a Universal Asynchronous Receiver/Transmitter (UART) as part of the peripheral set. The baud rate for the UART is normally generated by using a programmable counter to divide the frequency of some local oscillator. Depending on the frequency of the oscillator and the divisors, not all standard baud rates can be accurately generated. For reliable operation, a rule of thumb is to use a baud rate with a tolerance of ±5%.

The Programmable System-on-Chip™ (PSoC) microcontroller from Cypress MicroSystems is different to most in that it does not have fixed functional blocks. Rather, it has several digital and analog blocks that can be configured to generate the functions required on the chip. Despite that, the principles of this example are applicable to all microcontrollers although judicious selection of a suitable oscillator frequency can generate exact baud rates across the spectrum.

The PSoC microcontroller is a low cost part and in order to further the economy of the design, the PSoC has an onboard oscillator running at 24 MHz with an accuracy of ±2%. This oscillator can be divided by one or two prescalers, each of which can have a divisor of 1 through 16. The resultant of this division is connected to an 8-bit counter as shown in Figure 2-1. The baud rate generator is derived from the overflow of this 8-bit counter. The actual baud rate is 1/8 of the overflow rate and its value is derived from the formula:

$$(24 \times 10^6) / (24V1 * 24V2 * (N - 1) * 8),$$

where 24V1 is the first prescaler, 24V2 is the second prescaler, and N is the setting of the 8-bit counter. As with all microcontrollers, these internal settings are written to registers integral to the device. The complexity of the block configuration on the PSoC is masked by an elegant user interface "PSoC Designer," but that is beyond our concerns at the moment.

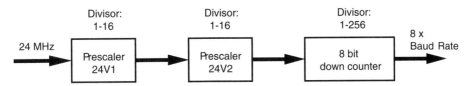

Figure 2-1: Divider chain in PSoC to generate a baud rate.

It is possible to start with a desired baud rate and work backwards to generate the desired divisors, but electronic design (or life, for that matter) isn't always so simple. There are always other issues. On the PSoC for instance, the prescaler outputs are used to drive other blocks and the chosen frequency needs to be compatible with those functions. The model is based on an actual example where the prescalers had been cast in stone in order to refresh a display at a constant rate. The available baud rates were subservient to this requirement. One of the advantages of computers is that they can execute repeated calculations in the blink of an eye, so this example takes a brute force approach. For a given set of prescalers, we will calculate all the possible baud rates.

Setup Workbook

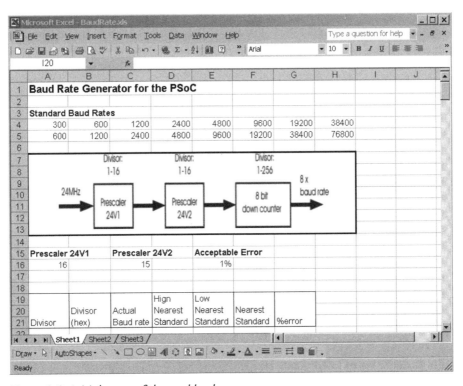

Figure 2-2: Initial setup of the workbook.

Figure 2-2 shows the initial approach at the spreadsheet. Before we analyze what is actually the intention of the workbook, we should look at some items of note.

In Parenthesis: *Merge Cells*

We saw in Example 1 how it was possible to maintain the visibility of headings when they occupy more room than one cell. A further technique to allow this is to block the cells where the text will appear and then following the sequence **Format Cells | Alignment** tab and checking the **Merge Cells** box. Where cells are merged, the first cell (leftmost in a horizontal merge, topmost in a vertical merge) contains the information. The other cells within the merged area cannot be used.

Of course, it is possible to combine several options as you see here in cells D19 to D21, where the cells are merged vertically and the word wrap feature is selected.

The title in cell A1 has merged 5 cells (see the accompanying "In Parenthesis: Merge Cells"). The same technique was used to format cells A3 to C3 as well as the sub headings in row 15. Cell A16 has been named to *PreV1*, cell C16 to *PreV2*, cell E16 has been named *AcceptableError* and has been formatted to percent. The block A4 to H5 has been named *StandardBaudRates*.

In Example 1, we saw that it is possible to insert a chart into a worksheet, so it is logical to presume that it is possible to insert a picture into the workbook. It could be used as documentation or for presentation. As with most Microsoft products, it is possible to use the sequence **Insert Picture | From File** and then select the format and picture as desired. It is possible to size, rotate and edit the image as normal. Unlike Word, however, there is no direct way to create a frame. This is easy to remedy. Block the cells that surround the picture and merge them, getting rid of the cell lines. Then, using the same block, create a border (**Format Cells | Border** tab) for the unified cell.

Let's consider the intended objectives for this model. The listing of the standard baud rates will be used as part of a lookup table. The prescalers can be set under the respective titles in row 16. We can also decide the acceptable error to be when choosing the baud rate, by changing the value in cell E16.

The table header (in rows 19-21) sets out what we are going to do. For each divisor in the 256 possible combinations, we will show the hexadecimal equivalent. Using the look up table of standard baud rates, we will find the nearest standard value less than the actual baud rate, and the nearest standard value greater than the actual baud rate. We will then find which is closest, and calculate the error.

Hexadecimal

In order to use the hexadecimal conversion, you must have enabled the add-ins as described in the beginning of the book or Excel will not recognize the function.

In Parenthesis: *Number Base Conversion*

It is possible to convert between the different number bases: decimal, binary hexadecimal, and octal using one of the following functions:

bin2hex(number, places)

bin2dec(number)

bin2oct(number, places)

dec2hex(number, places)

dec2bin(number, places)

dec2oct(number, places)

hex2dec(number)

hex2bin(number, places)

hex2oct(bin2hex (number, places)

oct2hex(number, places)

oct2dec(number)

oct2bin(number, places)

In above parameters "number" is the value in the source number base. "Places" is the number of digits that the answer will contain. It will "pad" the answer with zeroes for numbers that use less than the number of digits required by "places". If "places" is omitted, then only the minimum required number of digits to express the number are used.

Autofill all the numbers in column A, starting from the number 2 through to 255. In cell B22, add the formula:

=dec2hex(a22,2)

This results in a two digit hexadecimal number, but there is no real indication that it is hexadecimal. It is possible to add (concatenate) an "h" suffix (or a "0x" prefix) using string manipulation. Edit cell B22 to read:

=dec2hex(a22,2)&"h"

(or ="0x" & dec2hex(a22,2) for the alternative notation) and copy this cell from B22 to B275.

> **In Parenthesis:** *Split Screen*
>
> As we scroll down the workbook, the titles at the top of the columns disappear and there may be some confusion as to what the contents of a cell represent. This is very easy to solve by splitting the screen.
>
> It is possible to split horizontally, vertically or both. For a horizontal split, select a row by clicking on the row select button. In the example, we click on the row select button for row 22. Then go through the menu sequence **Window | Split**. There are now two windows and two copies of the same workbook. This way we can keep the heading in the one pane, and scroll up and down in the other. In order to do this vertically, we select a column with the column button and follow the same procedure. To split the window into four panes, we click on the cell that we want to be the nexus of the splits and once again follow the same procedure.
>
> To clear the split, we use the menu sequence **Window | Remove** split.
>
> It is also possible to lock a pane to disable any user shifting. In our case we may use it to hold the titles constant. To achieve this we click in the pane we want constant, and enter the menu sequence **Window | Freeze**. Now the bar that indicates the pane changes to a thin black line, and the selected pane remains constant while the cursor controls only affect the other pane.
>
> Unfreezing is as intuitive as unsplitting.

In cell C22, we enter the following formula:

$$=24E6/(8*pre24v1*pre24v2*(A22-1))$$

and then copy it to C23 through to C275. The E6 notation represents 10^6.

Lookup Tables

Before the advent of scientific calculators, lookup tables were all the rage for engineers, second only to slide rules. There were trigonometrical tables and logarithmic tables amongst others. Programmers use them to handle nonlinearities in sequences, high-speed access to data and many other applications. Excel provides this feature and it can be set up as a horizontal or a vertical lookup.

Let's consider the case of a horizontal lookup. The data is arranged in one or more rows as can be seen from our case for the standard baud rates. Using the actual baud rate I am going to search the first row of the table for the nearest standard value that is less than (or equal to) the actual baud rate. The second row is organized as the first row shifted to the left. For a value in the first row that is less than the actual baud rate, the corresponding value in the cell beneath the identified cell will be the standard baud rate above the actual baud rate.

In Parenthesis: *Lookup*

It is possible to lookup a value horizontally (or vertically) using the HLOOKUP (or VLOOKUP) function. The function takes the lookup value and searches the first row (or column) of the table. When a match is found (it can be an exact match or a value just less than the lookup value), a number is returned. This number is based on an offset number of rows (or columns) from the identified matching cell.

The format is:

HLOOKUP(lookup_value,table_array,row_index_num,range_lookup)

lookup_value is the number on which the search is based.

table_array defines the area of the table.

row_index_num is the offset from the matching cell. It cannot be negative and must be less than the total number of rows in to table.

range_lookup is a logical value. If set to TRUE or omitted, if an equal value is not found, the cell with the value closest to and less than the lookup_value is identified. If there are no more entries in the table, the highest value in the table is returned. If set to FALSE, an exact match must be found.

VLOOKUP follows the same format.

We recall that we had prenamed the lookup table as *StandardBaudRates*. We want cell E22 to contain the value below the calculated baud rate in C22, so we want to look in the first row of the table. The "index" value is therefore "1".

In cell E22, we enter the following formula:

> =hlookup(C22,StandardBaudRates,1,TRUE)

Now we want to identify the value higher than the actual baud rate, so we enter in cell D22 the formula:

> =hlookup(c22,StandardBaudRates,2,TRUE)

Note that the only difference is the "index" value. We block copy these two cells, D22, and E22, from D23 and E23 through to D275, E275. And, we can visually scan to see that it is performing as expected.

In order to decide whether the upper or lower standard baud rate is closer to the standard value, we need to subtract the actual baud rate from each of the standard values and compare the absolute values. This can be succinctly combined into an Excel formula that we enter in cell F22.

=IF((ABS(D22-C22)>ABS(E22-C22)),E22,D22)

Before we consider the next calculation, we should format the cells G22 through to G275 as a percentage with no digits after the decimal point, and then in cell G22 enter:

=(C22-F22)/F22

as a calculation of the error.

We block copy cells F22 and G22 through to row 275.

For a little formatting, block the whole table from A19 to G275 and format for a border and center justification. We should be looking at something like Figure 2-3.

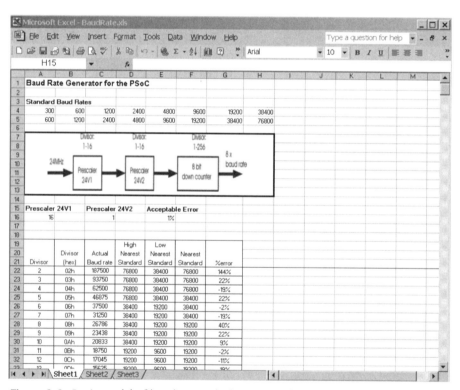

Figure 2-3: Basic model of baud rate calculation completed. Note the split and frozen screen line below row 21.

It is now possible to scan down and look for where the error is less than or equal to 1%. Of course, we can use the technique we developed in Example 1 where we add "<<<" to a column to the right to indicate where the acceptable values are. Let's do that. First, we format the text color of cells H22 to H275 to red. Then in cell H22 we enter:

$$=IF(ABS(G22)<=AcceptableError,"<<<","")$$

(remembering that "AcceptableError" had been prenamed) and copy this cell from H23 through to H275.

Conditional Formatting

We can actually go a step further. It is possible to change the formatting of a cell based on several conditions. As an example, when the error is less than the Acceptable Error and positive, the cell background should be yellow, the text red and in bold. If the error is negative and within the same range as before, the background color is green.

Block select cell H22 to cell H275. Select **Format** | **Conditional formatting...** and in the resulting dialog box we define the conditional for the formatting (as seen in the Conditional Formatting box on the lower left of Figure 2-4) and the desired format when the condition is met (in the Format Cells box in Figure 2-4) changing the font to red and bold under the **Font** tab and the cell shading under the **Patterns** tab.

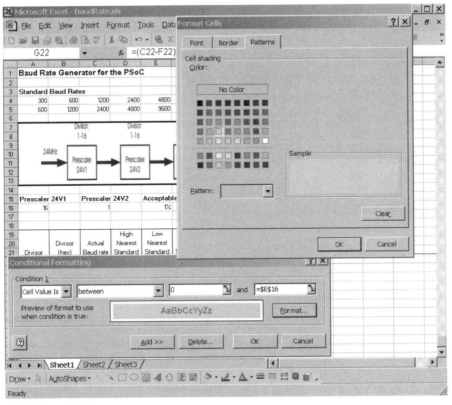

Figure 2-4: Defining the first condition of the cell format.

It is possible to have multiple conditions. Click on the **Add>>** button and define the second condition as in Figure 2-5. It is possible to have up to three conditions in total.

In Parenthesis: *Conditional Formatting*

Aside from the formatting provided in this example, you can achieve a number of effects.

It is possible to create alternating shaded rows like computer printouts used to have. First, we block the area that we want the effect. Next, we select the **Format** button in the same dialog box, click on the **Patterns** tab and choose the color that takes our fancy, and return via the **OK** button. Then we invoke the Conditional Formatting dialog box, selecting **Formula Is** from the drop-down menu on the left, and add the formula:

=row()=odd(row())

The **OK** button returns us to the workbook, and hopefully, the desired result.

Another thing we may want to do is to have formatting repeated every *N* rows. Perhaps a horizontal line every 8 rows. The process is very similar. First, we select the total area where we want to use the effect, and then get to the Conditional Formatting dialog box. We format for a border along the top of the cell only, (followed by the **OK**) and then change the action descriptor to **Formula Is** and add the following formula:

=mod(row(),8)=0

Click **OK** and we are done. Removing the "=0" at the end seems to have the opposite effect, removing the line every 8th row.

While it is not readily obvious that there is conditional formatting on a cell, it is possible to find cells with conditional formatting by using the **Edit | Go To | Special | Conditional Format**. Once there it is possible to select all cells, or to search for a cell with the same formatting as the current one.

When using this technique, we should keep in mind that, other than the visible formatting effect, there is no immediate indication (see "In Parenthesis: Conditional Formatting") that the cell has been conditionally formatted—something that could cause some problems when returning to a workbook after several years. Note there is no condition here for between or equal to so that it only highlights errors less than 1%. We could easily solve this by changing the acceptable value to 1.01% or some similar minor increment. You may notice that some 1% entries are not highlighted. This is because the rounding effect displays 1%, but the error is larger than this. In this case, the acceptable error could be set to 1.5%.

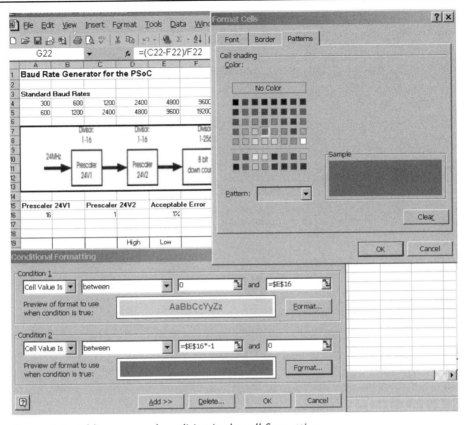

Figure 2-5: Adding a second condition in the cell formatting.

It is now possible to change the prescalers and pick out the possible baud rates. Figure 2-6 shows what the results would look like. It is possible to change the prescaler values and see what baud rates are feasible in the circumstances. The Excel file is titled *BaudRate.xls*.

Macro

OK, so it's still tedious to scroll down through 254 entries to find the ones that meet the criterion. We can create a macro to generate a table of just the valid entries. All macros in Excel are created in Visual Basic for Applications (VBA). So if we need to get a guide of how to write a macro or implement a particular feature, we can create a macro close to the desired action using the macro learn feature and then analyze it and either modify it or copy parts in order achieve our aims.

First, we will create a table header for a three-column table starting in cell K3. Since the titles are identical to the original table, it is simplest to copy each merged cell. What we want the macro to do is to scan down column H. If it detects a nonblank cell, then the pertinent information from the row is copied to the new table. The new table row is then formatted with a border and the text is centered.

We invoke the macro learn facility by the menu sequence **Tools I Macro I Record New Macro**. You will be presented with a dialog box to name the macro as in Figure 2-7.

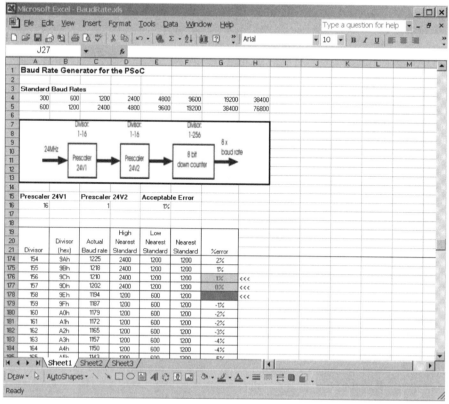

Figure 2-6: Completed basic model.

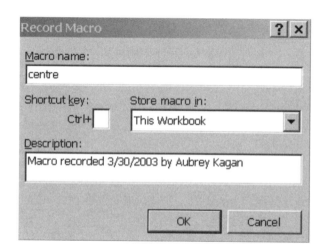

Figure 2-7:
Naming a macro
prior to "learning."

A small window box may pop up over the spreadsheet. (If it does not, it can be enabled by going through the sequence **View** | **Toolbars** | **Customize** | **Toolbars tab** | **Stop Recording**. But don't do this while recording the macro! You can always stop recording the macro by using the sequence **Tools** | **Macro** | **Stop Recording**.) The stop button will terminate the macro recording when clicked. Block any three cells in a row, say K6 to M6, right-click and select the sequence **Format cells** | **Alignment** tab and select the **Center** from the **Horizontal** drop-down selector. Stop recording the macro.

Now open VBA by going through the menus **Tools** | **Macro** | **Macros**, click on "center" and then click on the **Edit** button. It is also possible to get to Visual Basic by the menu options **Tools** | **Macro** | **Visual Basic Editor**. Once in the editor, go through the menu sequence **Tools** | **Macros** | **Edit**.

```
Sub centre()
' centre Macro
'
    Range("K6:M6").Select
    With Selection
        .HorizontalAlignment = xlCenter
        .VerticalAlignment = xlCenter
        .WrapText = False
        .Orientation = 0
        .AddIndent = False
        .IndentLevel = 0
        .ShrinkToFit = False
        .ReadingOrder = xlContext
        .MergeCells = False
    End With
End Sub
```

The above listing is the result. We would also like to know how to set up borders, so we follow the same idea in recording a macro that formats the cells for a border along all edges. We will call the procedure *frame*. The following listing is the result. If we follow either listing, we notice that the difference between this recorded sequence and what we are going to need is the selection of the cell block, that is, the line "Range("K6:M6").Select".

```
Sub frame()
    Range("K6:M6").Select
    Selection.Borders(xlDiagonalDown).LineStyle = xlNone
    Selection.Borders(xlDiagonalUp).LineStyle = xlNone
    With Selection.Borders(xlEdgeLeft)
        .LineStyle = xlContinuous
        .Weight = xlThin
        .ColorIndex = xlAutomatic
    End With
    With Selection.Borders(xlEdgeTop)
        .LineStyle = xlContinuous
```

```
            .Weight = xlThin
            .ColorIndex = xlAutomatic
        End With
        With Selection.Borders(xlEdgeBottom)
            .LineStyle = xlContinuous
            .Weight = xlThin
            .ColorIndex = xlAutomatic
        End With
        With Selection.Borders(xlEdgeRight)
            .LineStyle = xlContinuous
            .Weight = xlThin
            .ColorIndex = xlAutomatic
        End With
        With Selection.Borders(xlInsideVertical)
            .LineStyle = xlContinuous
            .Weight = xlThin
            .ColorIndex = xlAutomatic
        End With
    End Sub
```

Armed with this information, we create a new procedure titled *CompressPossibilties*. To do this click below the end of the macros we have created and type: "*sub CompressPossibilties*", followed by enter. VBA will automatically setup a blank procedure with the same name. Then, add the macro as follows. Large portions may be copied from the previously recorded macros.

```
Sub CompressPossibilties()
    Dim iVertInput As Integer
    Dim iVertOutput As Integer

    For iVertInput = 0 To 254 Step 1
        'set active cell to known position
        Range("H1").Select

        If (ActiveCell.Offset((iVertInput + 22), 0).Value <> "") Then
            ActiveCell.Offset((iVertOutput + 5), 3).Value = ActiveCell.Offset((iVertInput + 22), -7).Value
            'transferring cell in column A
            ActiveCell.Offset((iVertOutput + 5), 4).Value = ActiveCell.Offset((iVertInput + 22), -6).Value
            'transferring cell in column B
            ActiveCell.Offset((iVertOutput + 5), 5).Value = ActiveCell.Offset((iVertInput + 22), -3).Value
            'transferring cell in column C

            'formattingthe 3 cells with a border
            'this is derived from using the macro learn
            Range(ActiveCell.Offset((iVertOutput + 5), 3), ActiveCell.Offset((iVertOutput + 5), 5)).Select

                Selection.Borders(xlDiagonalDown).LineStyle = xlNone
                Selection.Borders(xlDiagonalUp).LineStyle = xlNone
```

```
With Selection.Borders(xlEdgeLeft)
   .LineStyle = xlContinuous
   .Weight = xlThin
   .ColorIndex = xlAutomatic
End With
With Selection.Borders(xlEdgeTop)
   .LineStyle = xlContinuous
   .Weight = xlThin
   .ColorIndex = xlAutomatic
End With
With Selection.Borders(xlEdgeBottom)
   .LineStyle = xlContinuous
   .Weight = xlThin
   .ColorIndex = xlAutomatic
End With
With Selection.Borders(xlEdgeRight)
   .LineStyle = xlContinuous
   .Weight = xlThin
   .ColorIndex = xlAutomatic
End With
With Selection.Borders(xlInsideVertical)
   .LineStyle = xlContinuous
   .Weight = xlThin
   .ColorIndex = xlAutomatic
End With

With Selection
'this will center the entries in the block

   .HorizontalAlignment = xlCenter
   .VerticalAlignment = xlCenter
   .WrapText = False
   .Orientation = 0
   .AddIndent = False
   .IndentLevel = 0
   .ShrinkToFit = False
   .ReadingOrder = xlContext
   .MergeCells = False
End With
iVertOutput = iVertOutput + 1
'bumping on to next row in output

   End If
Next

'return active cell to start of table
Range("A22").Select
End Sub
```

The CompressPossibilties macro searches for nonblank entries in column H from rows 22 to 275, and when the nonblank cell is detected, it will copy the relevant information to a concise table on the right of the worksheet. In the VBA code in Excel, one technique to accessing cells in a worksheet is to select a particular cell as the reference position and then manipulate all the other cells relative to that one. I arbitrarily chose cell H1 as can be seen from the *Range("H1").Select* statement.

Running this macro (**Tools** | **Macro** | **Macros**, click on CompressPossibilties from the list and click on **Run**) will generate a table as seen on the upper right-hand side of the workbook in Figure 2-8.

Once this the macro has been coded, we can delete the "center" and "frame" macros since we no longer need them, but I have left them in "BaudRateWithMacro.xls" as a reference.

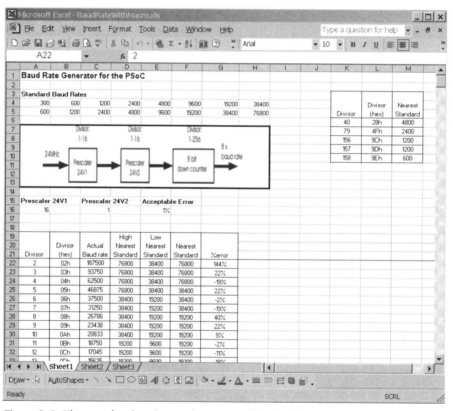

Figure 2-8: The complete baud rate selection workbook. The heading K3 to M3 has been set up as part of the workbook. Each row beneath this title is added by the macro replete with center alignment and border formatting.

EXAMPLE 3

Mean Time Between Failures (MTBF)

Model Description

The difficulty of reworking products using surface-mount devices and the cost of labor have turned the concept of Mean Time Between Failures (MTBF) into a bit of a misnomer since the concept was derived for a unit that is repairable. Nevertheless, MTBF remains as the most quoted measure of equipment reliability.

The probability that a piece of equipment will experience no failures in time t (that is, the reliability) can be derived from the equation:

Reliability = $e^{-\lambda t}$

As usual in engineering, e represents the base of Naperian logarithms (2.718). λ is a constant representing the failure rate. The MTBF is the inverse of the failure rate:

MTBF = $1/\lambda$

There are several methods of calculating MTBF for a piece of equipment. The most common is based on the failure rate of each component, the number of components and the environmental stresses on the components.

For each device in a piece of equipment, the steady-state failure rate is given by:

$\lambda_{SSi} = \lambda_{Gi}\,\Pi_{Qi}\,\Pi_{Si}\,\Pi_{Ti}$

where:

   λ_{SSi} is the Steady-State Failure Rate for the device under consideration.

   λ_{Gi} is the Generic Failure Rate for the device under consideration.

   Π_{Qi} is the Quality Factor for the device under consideration.

   Π_{Si} is the Stress Factor for the device under consideration.

   Π_{Ti} is the Temperature Factor for the device under consideration under normal operating conditions.

Based on each individual component, it is then possible to derive the overall steady-state failure rate by summing the failure rate for each component part and applying a unit environmental factor

$$\lambda_{SS} = \Pi_E \, \Sigma \, \lambda_{SSi}$$

where:

the sum is from $i = 1$ the total number of devices in the unit.

λ_{SS} is the overall steady-state failure rate.

Π_E is the unit Environment Factor.

Once this is done, as we have seen MTBF $= 1/\lambda_{SS}$

The information required to calculate this quantity is derived from statistical analysis of observed data. It is normally presented in the form of tables and so it is not too difficult to create an Excel model to make this calculation.

It should be noted that the numbers that I have used are derived from a dated document, *Bellcore Technical Reference TR-NWT-000332*, Issue 3, September 1990. Use these tables with caution since they a most likely to have been updated in line with the changes in the electronics industry in packaging and transistor density. In addition, different organizations (especially the military) may use different specifications in deriving their MTBF requirements.

Factors

Figure 3-1, Figure 3-2, Figure 3-3, and Figure 3-4 represent the different factors (denoted by Π_X above) entered onto different sheets in the workbook. It is not possible to enter the title Π_X directly onto the tab, but it can be done by writing the text (using **Insert** | **Symbol**) into a cell and copying from the cell to the tab. The subscript formatting is not copied.

In Parenthesis: *Multiple Worksheets*

By default, Excel initiates a workbook (based on a template) with three worksheets named *Sheet1*, *Sheet2* and *Sheet3*. (Incidentally, it is possible to create new templates and even change the default template that Excel uses.)

To change the name of the worksheet, change the color of the tab, insert or delete a tab, the secret is similar to many other Windows applications. Right-click on the tab and follow the selections.

By using the right-click, reordering or copying tabs is possible. It is also possible to move a worksheet tab by clicking and dragging the tab until a small black triangular indicator (just above the tabs) shows where the result of the move will be.

| A8 | | ▼ | | *fx* | | |

	A	B	C	D	E	F
1	Device Quality Factor					
2	Quality Level	IC		Discrete		Other
3		Hermetic	Non-Hermetic	Hermetic	Plastic	
4	1	1.5	1.8	1.5	1.8	1.5
5	2	1.0	1.0	1.0	1.0	1.0
6	3	0.5	0.5	0.5	0.5	0.5

Figure 3-1: Device Quality Factor. When we reference this table based on information in the Bill of Materials, column B will be considered column 1, Column C as 2, and so forth. This is implicit in the naming of the range PI_Q for cells B4 to F6.

	A	B	C	D	E	F	G	H	I	J	K	L	M
1	Electrical Stress Curve												
2	% stress	A	B	C	D	E	F	G	H	I	J	K	L
3	0%	0.8	0.7	0.6	0.5	0.4	0.3	0.2	0.2	0.2	0.1	1.0	1.0
4	10%	0.8	0.7	0.6	0.5	0.4	0.3	0.2	0.2	0.2	0.1	1.0	1.0
5	20%	0.8	0.8	0.7	0.6	0.5	0.4	0.3	0.3	0.3	0.2	1.0	1.0
6	30%	0.9	0.8	0.8	0.7	0.6	0.6	0.5	0.4	0.4	0.3	1.0	1.0
7	40%	0.9	0.9	0.9	0.8	0.8	0.7	0.7	0.7	0.6	0.6	1.0	1.0
8	50%	1.0	1.0	1.0	1.0	1.0	1.0	1.0	1.0	1.0	1.0	1.0	1.0
9	60%	1.1	1.1	1.1	1.2	1.3	1.3	1.4	1.5	1.6	1.8	1.1	1.0
10	70%	1.1	1.2	1.3	1.5	1.6	1.8	2.0	2.3	2.5	3.3	1.1	1.0
11	80%	1.2	1.3	1.5	1.8	2.1	2.4	2.9	3.4	4.0	5.9	1.2	1.0
12	90%	1.3	1.4	1.7	2.1	2.6	3.2	4.1	5.2	6.3	10.6	1.3	1.0

Figure 3-2: Electrical Stress Factors. Column A is formatted as percentage and the other data is formatted to one decimal place. Everything is formatted for center alignment. Note the additional Excel column M (entitled "L") to allow for devices that do not require the electrical stress factor. There is also a 0% row to allow the use of the VLOOKUP function. I took the liberty of duplicating the 10% row for the new row. The range A3 to M12 is named PI_S.

	A	B	C	D	E	F	G	H	I	J	K
1	Temperature Factors										
2	Operating	Temperature Stress Curve									
3	Temperature (∘C)	1	2	3	4	5	6	7	8	9	10
4	0	1.0	0.9	0.9	0.8	0.7	0.7	0.6	0.6	0.5	0.4
5	30	1.0	0.9	0.9	0.8	0.7	0.7	0.6	0.6	0.5	0.4
6	31	1.0	0.9	0.9	0.8	0.7	0.7	0.6	0.6	0.5	0.5
7	32	1.0	0.9	0.9	0.8	0.8	0.7	0.6	0.6	0.6	0.5
8	33	1.0	0.9	0.9	0.9	0.8	0.7	0.7	0.7	0.6	0.6
9	34	1.0	0.9	0.9	0.9	0.8	0.8	0.7	0.7	0.7	0.6
10	35	1.0	1.0	0.9	0.9	0.9	0.8	0.8	0.8	0.7	0.7
11	36	1.0	1.0	0.9	0.9	0.9	0.8	0.8	0.8	0.8	0.7
12	37	1.0	1.0	1.0	0.9	0.9	0.9	0.9	0.9	0.8	0.8
13	38	1.0	1.0	1.0	1.0	0.9	0.9	0.9	0.9	0.9	0.8
14	39	1.0	1.0	1.0	1.0	1.0	1.0	1.0	0.9	0.9	0.9
15	40	1.0	1.0	1.0	1.0	1.0	1.0	1.0	1.0	1.0	1.0
16	41	1.0	1.0	1.0	1.0	1.0	1.0	1.0	1.1	1.1	1.1
17	42	1.0	1.0	1.0	1.0	1.1	1.1	1.1	1.1	1.1	1.2
18	43	1.0	1.0	1.0	1.1	1.1	1.1	1.2	1.2	1.2	1.3
19	44	1.0	1.0	1.1	1.1	1.1	1.2	1.2	1.2	1.3	1.4
20	45	1.0	1.0	1.1	1.1	1.2	1.2	1.3	1.3	1.4	1.5
21	46	1.0	1.1	1.1	1.1	1.2	1.3	1.3	1.4	1.5	1.6
22	47	1.0	1.1	1.1	1.2	1.3	1.3	1.4	1.4	1.6	1.8
23	48	1.0	1.1	1.1	1.2	1.3	1.4	1.4	1.5	1.7	1.9
24	49	1.0	1.1	1.1	1.2	1.3	1.4	1.5	1.6	1.8	2.1
25	50	1.0	1.1	1.1	1.2	1.4	1.5	1.6	1.7	1.9	2.2
26	51	1.0	1.1	1.2	1.3	1.4	1.6	1.7	1.8	2.0	2.4
27	52	1.0	1.1	1.2	1.3	1.5	1.6	1.7	1.9	2.2	2.6
28	53	1.0	1.1	1.2	1.3	1.5	1.7	1.8	1.9	2.3	2.8
29	54	1.0	1.1	1.2	1.4	1.6	1.7	1.9	2.0	2.4	3.0
30	55	1.0	1.1	1.2	1.4	1.6	1.8	2.0	2.1	2.6	3.3
31	56	1.0	1.2	1.2	1.4	1.7	1.9	2.1	2.3	2.8	3.5
32	57	1.1	1.2	1.3	1.4	1.7	2.0	2.2	2.4	2.9	3.8
33	58	1.1	1.2	1.3	1.5	1.8	2.0	2.2	2.5	3.1	4.1

H ◀ ▶ H \ BOM / ΠQ / ΠS \ ΠT /

Figure 3-3: Temperature Factors. I added a row for 0 degrees in order to use the VLOOKUP function. The range A4 to K57 is named PI_T.

	A	B	C
1	Environment	E SYMBOL	Π_E
2	Ground, Benign	GB	1.0
3	Ground, Fixed	GF	1.5
4	Ground, Mobile	GM	5.0
5			

H ◀ ▶ H \ BOM / ΠQ / ΠS / ΠT \ ΠE /

Figure 3-4: Environmental Factors. The range B2 to C4 is named PI_E.

Bill of Material

The first step in building the model is entering a Bill of Material into Excel. There is the old-fashioned way of entering each part, key press by key press, but for anything more than a few parts, this can be a laborious process. If you are lucky enough to use a product like Parts & Vendors, it will have a direct Excel export facility and you can load the Bill of Materials in a few mouse clicks. There is an alternative file format that Excel can load, known as the "comma delimited" format that uses a "csv" file suffix. If it is possible to have the Bill of Material formatted in this way, entry of the parts should be a breeze.

In Parenthesis: *Comma Delimited Files*

In this dated but simple file format, text and information is saved in a text file in the same way that the data is seen on the screen. Separation between the entries is by means of a delimiter. Originally this was a comma, but Excel does allow for other delimiters as well. It is traditional to have the column titles as the first "row" of the file.

To create a comma delimited file to experiment with, create a worksheet in Excel with different text and numerical entries. Then using the **File | Save as...** sequence, select the **CSV (comma delimited) (*.csv)** filetype. It is possible to open this with a text processor and analyze the file contents. Importing the data can be done either by opening a *.csv file, or going through the sequence **Data | Import External Data | Import data...**

If we open the file, because of the simplicity of the format, we will be left with a file with a single worksheet beginning at cell A1. This can easily be manipulated, but with a large file, it may be cumbersome. Importing data allows the data to be placed anywhere in the workbook.

This single-sheet effect means that it is a little more difficult to build a model that can be reused with different Bills of Material. For more information on how to approach this, look at the Microsoft Knowledge Base Article 213816 titled, *XL2000: Macro to import a text file into an existing workbook*.

There is one more possible approach to use to save entry time if we are likely to be doing an MTBF analysis on a regular basis. Each device has a failure rate, quality factor and a maximum rating (used in the electrical stress calculation) associated with it. In addition, inherent to the failure rate for each device are two "curve" types (reduced to tables) for the electrical stress factor and the temperature stress factor. For instance, there is a different temperature stress curve for transistors and resistors. In fact, it is possible even to have different stress curves within a classification like "integrated circuits." Each device also needs a category for the Device Quality Factor calculation.

Trying to create and maintain a lookup table of failure rates for individual components is probably a guarantee of rather boring lifetime employment, so my suggestion is to try and get five extra fields in the part number database. When a part number is generated, these five fields should be filled out based on the latest available data. Parts & Vendors has the capability to have up to five user fields (in P&V under **Edit | Options** and select the **User** tab). It is possible to get by with a single field since Excel could parse the field into the five columns of information.

The five fields required for each part in the Bill of Materials are: the Device Failure Rate λ, the Temperature Stress Curve that should be used, the Electrical Stress Curve that should be used, the Device Quality Factor Π_Q, and the maximum electrical stress permitted for the device.

The Device Failure Rate λ, is the number of failures in 10^9 hours (also known as FIT). The actual numbers are derived from published tables provided with the specification or some other authority.

Each device has a Temperature Stress curve associated with it. In Figure 3-3, the "curve" is represented by a column labeled "1" to "10". The actual factor used Π_T, is the number that is at the intersection of the Curve column and the temperature of operation of the device.

In a similar manner, each device has an Electrical Stress Curve associated with it. In Figure 3-2, the "curve" is represented by a column labeled "A" to "L". The actual factor used Π_S, is the number that is at the intersection of the Curve column and the percentage electrical stress that device is subjected to. For most devices, the electrical stress is the ratio of actual power dissipated by the part to the maximum power dissipation allowed. Some components like capacitors and relays use different ratios. For capacitors, the electrical stress is the ratio of voltage across the capacitor to the maximum allowed voltage.

Every device has a quality factor depending on the device type and the packaging technology, as can be seen from the Device Quality Factors table Π_Q in Figure 3-1. The columns are numbered 1–5 for simplicity, with 1 corresponding to the Integrated Circuits "Hermetic" column.

In addition, when undertaking a MTBF analysis, there are some global variables that must be decided by the user. These are the ambient temperature, the Quality Level, and the Environmental Condition Factor Π_E.

In Parenthesis: *Comments*

A comment may be added to any cell in a worksheet. When the cursor hovers above the cell, a pop-up message appears. This message can be used to prompt the user for an input or to remind the programmer what his original intention was. A cell that contains a comment has a small red triangle in the top right-hand corner.

Adding a comment is simple. Right-click on a cell, and then select **Insert Comment** from the pop-up menu. Text can be pasted into the comment box. Right-clicking on a cell containing a comment will allow the text to be edited or deleted.

The Quality Level is defined as part of the overall product specification. For instance, it could be specified by the customer or by the intended environment for the product. The Quality Level defines the row that will be used in the table if Figure 3-1.

The Environmental Condition is derived from the table in Figure 3-4. It is defined as part of the product specification. We need to choose one of three possibilities based on the type of ground the system employs.

The "Maximum Electrical Stress" column carries no units, so it is simply the number that represents the maximum value to be used in the ratio when calculating the electrical stress.

With or without the Bill of Material importation, this information must appear in the worksheet in order to calculate the MTBF. The data in the tables presented here is used as a guide only. You must obtain the data for the parts that you use from the latest tables.

The initial data input is shown in Figure 3-5. Because of the size of the worksheet, a shifted view is shown in Figure 3-6.

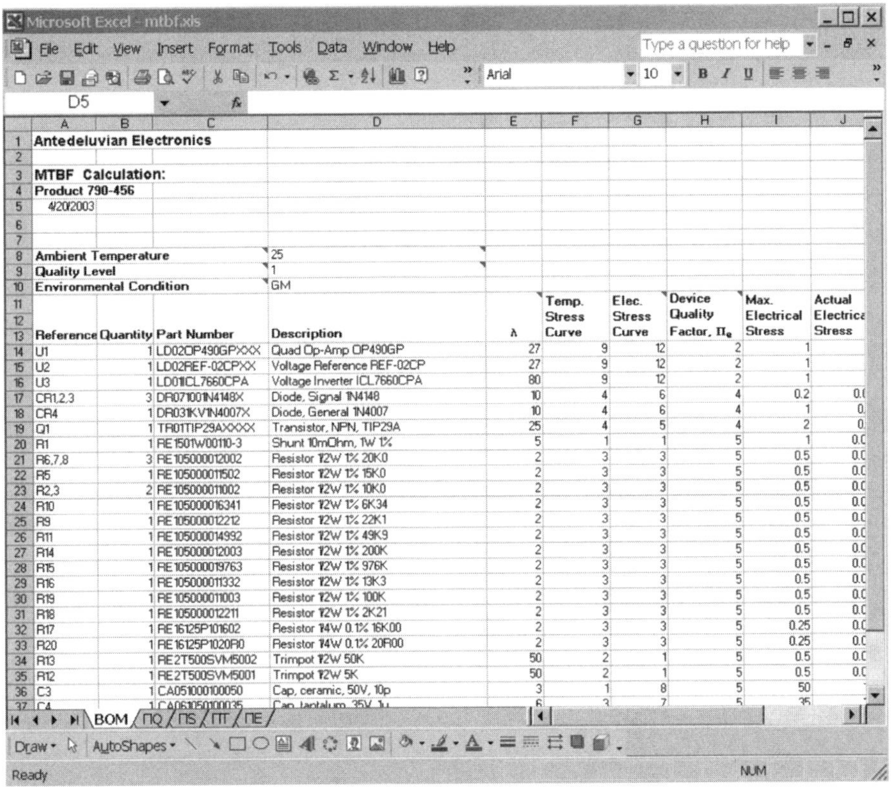

Antedeluvian Electronics

MTBF Calculation:
Product 790-456
4/20/2003

Ambient Temperature 25
Quality Level 1
Environmental Condition GM

Reference	Quantity	Part Number	Description	λ	Temp. Stress Curve	Elec. Stress Curve	Device Quality Factor, Π_e	Max. Electrical Stress	Actual Electrica
U1	1	LD02OP490GPXXX	Quad Op-Amp OP490GP	27	9	12	2	1	
U2	1	LD02REF-02CPXX	Voltage Reference REF-02CP	27	9	12	2	1	
U3	1	LD01ICL7660CPA	Voltage Inverter ICL7660CPA	80	9	12	2	1	
CR1,2,3	3	DR07100IN4148X	Diode, Signal 1N4148	10	4	6	4	0.2	0.(
CR4	1	DR031KV1N4007X	Diode, General 1N4007	10	4	6	4	1	0.
Q1	1	TR01TIP29AXXXX	Transistor, NPN, TIP29A	25	4	5	4	2	0.
R1	1	RE1501W00110-3	Shunt 10mOhm, 1W 1%	5	1	1	5	1	0.0
R6,7,8	3	RE105000012002	Resistor 1/2W 1% 20K0	2	3	3	5	0.5	0.0
R5	1	RE105000011502	Resistor 1/2W 1% 15K0	2	3	3	5	0.5	0.0
R2,3	2	RE105000011002	Resistor 1/2W 1% 10K0	2	3	3	5	0.5	0.0
R10	1	RE105000016341	Resistor 1/2W 1% 6K34	2	3	3	5	0.5	0.0
R9	1	RE105000012212	Resistor 1/2W 1% 22K1	2	3	3	5	0.5	0.0
R11	1	RE105000014992	Resistor 1/2W 1% 49K9	2	3	3	5	0.5	0.0
R14	1	RE105000012003	Resistor 1/2W 1% 200K	2	3	3	5	0.5	0.0
R15	1	RE105000019763	Resistor 1/2W 1% 976K	2	3	3	5	0.5	0.0
R16	1	RE105000011332	Resistor 1/2W 1% 13K3	2	3	3	5	0.5	0.0
R19	1	RE105000011003	Resistor 1/2W 1% 100K	2	3	3	5	0.5	0.0
R18	1	RE105000012211	Resistor 1/2W 1% 2K21	2	3	3	5	0.5	0.0
R17	1	RE16125P101602	Resistor 1/4W 0.1% 16K00	2	3	3	5	0.25	0.0
R20	1	RE16125P1020R0	Resistor 1/4W 0.1% 20R00	2	3	3	5	0.25	0.0
R13	1	RE2T500SVM5002	Trimpot 1/2W 50K	50	2	1	5	0.5	0.0
R12	1	RE2T500SVM5001	Trimpot 1/2W 5K	50	2	1	5	0.5	0.0
C3	1	CA051000100050	Cap, ceramic, 50V, 10p	3	1	8	5	50	
C4	1	CA061050100035	Cap, tantalum, 35V, 1u	6	3	7	5	35	

Figure 3-5: Initial setup of bill of materials.

This information for each component in the Bill of Materials in the worksheet is filled in the respective columns E to I, or if you have managed to configure the Bill of Material database, it will be imported directly. Up to this point, all the factors discussed for these columns are fixed and can be permanently associated with the component, hence the advantage of embedding them in the database.

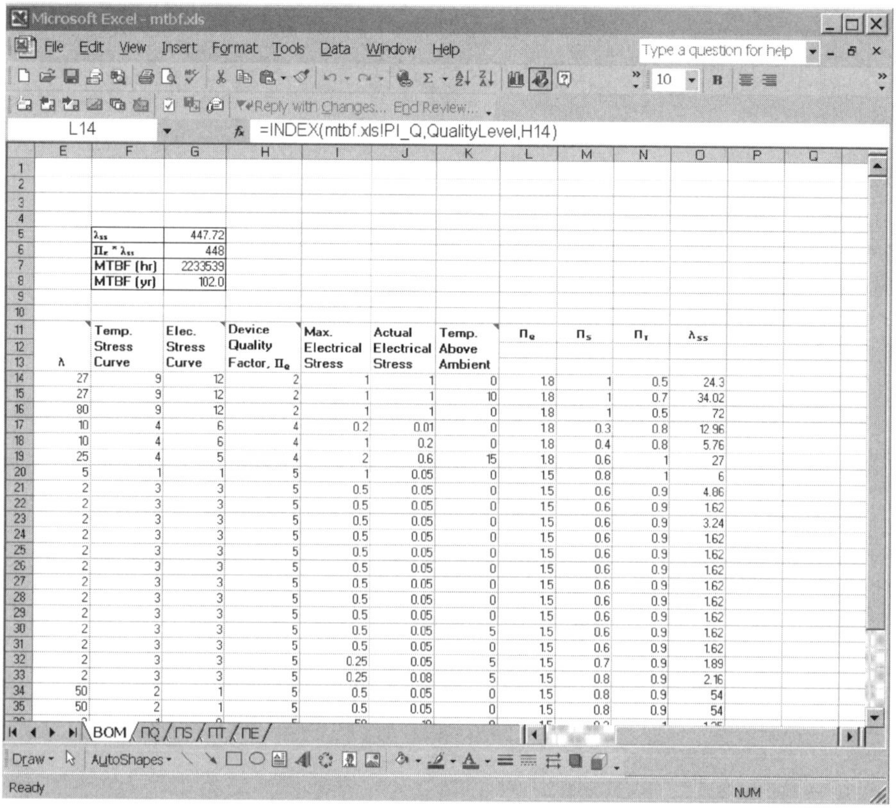

Figure 3-6: View of the right-hand extents of the worksheet. Columns E to I contain data that is constant for a given device. Columns J and K have data derived from a working unit.

From a working unit, measurements must be taken to determine the stress levels and these are entered in the Actual Electrical Stress column. In real-life, measuring the power dissipation in a resistor (or similar device) is done by measuring the voltage across the resistor and then using the power calculation (V^2/R) to generate the power. If we figure out what the maximum V^2 is for a device (for a ½W resistor it would be P*R or 0.5*R) and use it as the maximum stress entry, we can then enter the voltage measured in the actual stress cell and use Excel to calculate the square. I will continue with this thread a little later.

Also, as a result of self-heating of onboard components and the thermal gradient away from the heat source, the temperature of some components may be above ambient. This too should be measured and added in column "Temp. Above Ambient".

In cell A5, I entered the formula:

=*Today()*

so that the current date is always shown, which is useful in identifying the latest version of the workbook when there are several copies of the document that have been printed.

On Sheet "BOM", I have named cell D8 *AmbientTemperature*, cell D9 *QualityLevel*, and cell D10 *EnvCondition*. On sheet "ΠQ", I have named the range B4:F6 *PI_Q*, on sheet "ΠS" range A3:L12 *PI_S*, on sheet "ΠT" range A4:K57 *PI_T*, and on sheet "ΠE" B2:C4 *PI_E*.

Calculating the Quality Factor

To start the initial model design, we enter the titles of the three columns Π_Q, Π_S, and Π_T (see Figure 3-6). I intend to calculate these individually to allow debugging and for ease of explanation.

To find the desired quality factor we need to know the Quality Level, which is the number contained in cell D9 (on the BOM sheet, named *QualityLevel*). This determines which row the information is in on worksheet ΠQ. The device quality factor (BOM sheet, cell H14 for the first component) determines which column. We just need to use this information for the INDEX function.

> **In Parenthesis:** *INDEX*
>
> The INDEX function returns a value from a table based on its row and column identifiers. The index function has the syntax:
>
> =INDEX(**array**,row_num,column_num)
>
> where the array is the lookup matrix, and the row_num and column_num are self explanatory.

If we click on cell L14 and enter the text:

> =*index(*

Excel will respond with a prompt as to what syntax format to follow. The table we want is on another worksheet and to compound the problem it is named with a character that Excel won't accept in the data entry. The simple solution is to click on the Π_Q tab and the worksheet name is automatically entered followed by an exclamation point as the delimiter of the sheet. The data entry so far would appear as:

> =*index(ΠQ!*

Add the name of the range in the table PI_Q and add a comma so that the entry appears as:

> =*index(ΠQ!PI_Q,*

and then click on the BOM tab. Add the row number by typing the name "QualityLevel", add a comma and type "h14". Close the parenthesis so that you have:

> =*index(ΠQ!PI_Q,BOM!qualitylevel,h14)*

<Enter> will terminate the entry and return a value of 1.8, which is the quality factor for a nonhermetic integrated circuit at quality level 1.

Excel will use defaults if the page and workbook entries are not specified. Rather like private and public variables in programming it starts with the local and expands outwards. For instance, we didn't need the *BOM!* in the formula entry. It just got placed there when we clicked on the worksheet tab. Of course we could enter the text directly (with the exception of special characters like Π) without following the prompts of the data input.

We now copy this cell to all the necessary cells in the Π_Q column.

Calculate Electrical Stress Factor

The electrical stress is defined for each row as the contents of column J (actual stress) divided by the contents of column I (maximum stress). Since this does not work out to an integer value, we will need to use a lookup function that allows for a noninteger offset. We will need the VLOOKUP function. Its syntax is:

=VLOOKUP(lookup_value,table_array,col_index_num,range_lookup)

The lookup value is the number used for the lookup, and for row 14 would be the result of the division J14/I14. The division can be used as the parameter directly. The table array is located on sheet "ΠS" and is named *PI_S*.

Let's put this together. Click on cell M14 and enter:

=vlookup((J14/I14),

Click on the ΠS tab and type in PI_S followed by a comma so that the entry is:

=vlookup((J14/I14), ΠS!PI_S,

and then click on the **BOM** tab (although not really necessary), add *G14+1* and close the parenthesis and **<Enter>**. So that the full entry is:

=vlookup((J14/I14), ΠS!PI_S,BOM!G14+1)

The +1 is to allow for the fact that the first column in the range of the table PI_S is the percentages. As an aside, the cell description in the above line is BOM!G14. If we wanted to add parenthesis to the BOM!G14+1, it would have to be (BOM!G14+1) and *not* BOM!(G14+1).

This entry returns a value of 1 corresponding to the electrical stress of an integrated circuit. Copy the cell M14 to M15 through to M41.

The reason that I added a 0% row is that the VLOOKUP function needs to find a number in the column that is less than the lookup_value. If it does not, it will return an error message. Clearly it is possible to have a stress level less than 10% (but more than 0%) so a row of 0% was added.

As I mentioned earlier, in order to save time, instead of entering resistor power we could enter resistor voltage. We could then modify the above line to be:

$=vlookup(if(G14=3,(J14^2)/I14,(J14/I14)), \Pi S!PI\_S,BOM!G14+1)$

so that if it is a resistor, the value of J14 would be squared. In an attempt to keep this simple, I have not implemented this.

In Parenthesis: *Table Functions*

There are several functions included with Excel that relate to tabular lookup. For detailed explanations use the Excel Help function, but as a simplified summary of what to consider, here is a list and a brief explanation.

HLOOKUP: in a table, search a row for a particular value (or between values) and then using the identified cell as a reference point, return the value from the cell in the same column, but offset by a defined number of rows.

VLOOKUP: in a table, search a column for a particular value (or between values) and then using the identified cell as a reference point, return the value from the cell in the same row, but offset by a defined number of columns.

LOOKUP: is very similar to HLOOKUP and VLOOKUP, only works with a single input range (column or row) and returns a value from a single output range (column or row).

MATCH: has the same approach as LOOKUP, but instead of returning the value of the addressed cell, it returns the position in the lookup array.

INDEX: returns the value from the intersection of a particular column and a particular row in a range.

Other associated functions are:

COLUMN: returns a number associated with the column of a cell or range of cells (A=1, B=2, and so forth).

COLUMNS: returns the number of columns in a range.

ROW: returns a number associated with the row of a cell or range of cells.

ROWS: returns the number of rows in a range.

TRANSPOSE: will return a horizontal range as a vertical range, or a vertical range as a horizontal range.

ADDRESS: returns text with the cell co-ordinates and can be relative to current cell use relative and absolute addressing.

AREAS: returns the number of contiguous cells in a range.

CHOOSE: returns a value from a list using an index

INDIRECT: will be familiar to users of assembly or C code. Excel takes the contents of a cell and uses it as the cell co-ordinates for a fetch.

OFFSET: accesses a cell or range of cells, offset from a cell or range.

The composite temperature to use for the Temperature Stress Curve is the ambient temperature plus the value in the associated row of column K, "Temperature Above Ambient". Since the table in the specification only begins at 30°C, we need to consider what will happen with an operating temperature below that. We could use the INDEX function because we are working with integer temperatures, but what will happen to sub 30°C values? Rather than complicate the INDEX function by adding an IF statement, we will resort back to the VLOOKUP, but that means that there needs to be an entry row below 30°C. If there is a likelihood of subzero temperatures, the minimum value should be entered here. I opted for 0°C.

The lookup_value parameter for the VLOOKUP is the sum of the cell named *AmbientTemperature* and the relevant cell in column K. The result should be:

   =VLOOKUP((AmbientTemperature+K14),ΠT!PI_T,BOM!F14+1)

This cell should be copied to L15 through to L41.

Calculation of λ_G

For each device type, the Steady State Failure Rate is the Device Failure Rate multiplied by the Device Quality Factor, the Electrical Stress Factor and the Temperature Stress Factor. This is true for each occurrence of the device so we need to multiply by the number of identical devices used.

Click on cell O14 and enter:

   *=B14*E14*L14*M14*N14*

and copy as before through to O41. If similar devices are subjected to different stresses they should be treated individually.

The total Steady-State Failure Rate is the sum of the cells from O14 to O41. Before I generalize the model, let me show you a quick way to enter the sum. Block from cell O14 to O43. Then click on the Σ button on the toolbar, and presto—it is done! Now, click on the undo button to get rid of this. I would like to place the results of the workbook above the Bill of Materials entries. My reasoning is that each Bill of Materials is going to have a different number of components and I would like to save entering the SUM every time.

I created a result table in cells F5 to G8. In G5, I entered the formula:

   =sum(O14:O300)

Summing blank cells does not affect the total so I chose an arbitrarily large number. We will see in later examples how to use the COUNT and COUNTA functions to determine how many rows actually exist.

Cell G6 must contain the product of λ_{ss} (cell G5) and the environmental conditions based on cell D10 (named *EnvCondition*). It contains:

=$G$5*(VLOOKUP(EnvCondition,ΠE!PI_E,2,FALSE))

Note the use of the FALSE parameter to indicate that the lookup must find the exact entry.

Now the MTBF in hours is the inverse of cell G6 multiplied by 10^9. It is a trivial task to divide down to generate years. In cell G7, I entered the formula:

=1E9/g6

and in G8:

=$G$7/(60*365)

Figure 3-7 shows the result of the calculations and the completed model. One of the things to consider is the further generalization of the model. Depending on the file format, it is quite possible that importing the Bill of Material will overwrite the formulas. Aside from the approach given in the Microsoft article discussed previously, a possible solution is to do the calculations on a separate worksheet, but the problem with that is we lack the immediacy of working alongside the data. What we can do is go to each of the cells L14, M14, N14 and O14 and place an apostrophe ' in front of the line turning the formulas into text. Copy these cells to another worksheet and paste them somewhere. Now once the bill of materials is loaded, these cells can be copied back and the apostrophes removed.

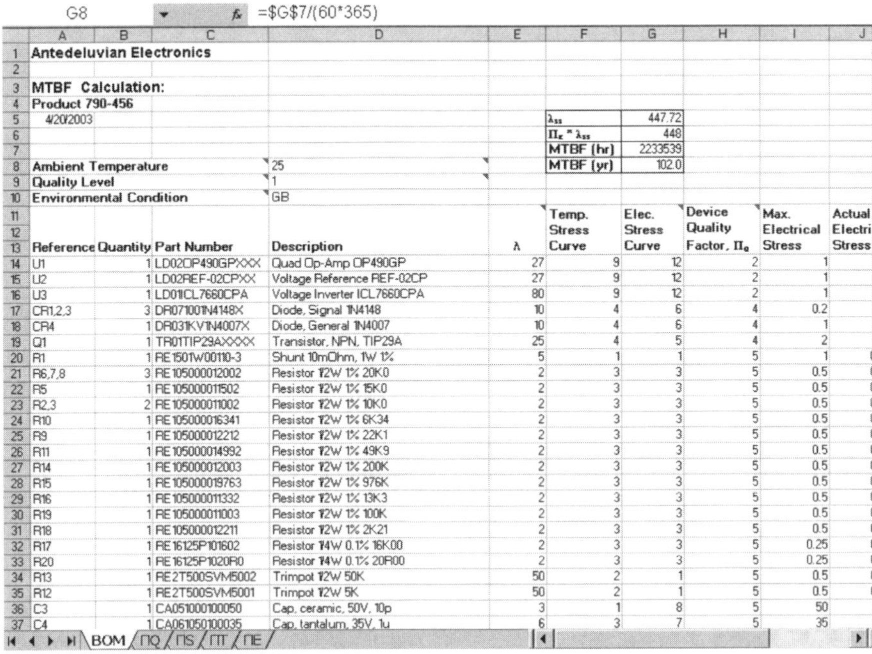

Figure 3-7: Model to generate MTBF. Note the summary block in cells F5:G8.

Scenario

Now that we have completed our model, we can play "what if" by changing the ambient temperature or consider the implication on the MTBF of the added cost of choosing a higher quality level or even a change in the harshness of the environment. Each time we change these parameters, the MTBF will be updated. Unless we print out each case, we are forced to remember just what the result was. Now in this instance this may be easy enough, but in other models the result may be a lot more complex, so let's use the opportunity to investigate the Scenario Manager tool. As with all complex models, it is probably a good idea to create a copy of the original file, just in case Murphy was right. (Actually, Murphy was an optimist!)

Select the sequence **Tools | Scenarios...** and you will be faced with the dialog in Figure 3-8 since no scenarios have been yet been defined.

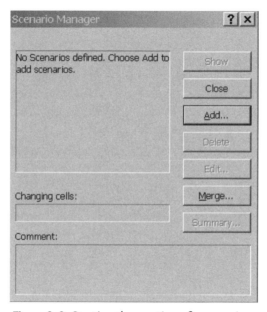

Figure 3-8: Starting the creation of a scenario.

Click on the Add button and you will be presented with a further dialog similar to Figure 3-9. The scenario manager allows you to specify which cells are going to change in each scenario. Initially, the active cell appears in the **Changing cells:** box. If this cell is not one of the cells that will change, we must delete the entry in the box. The cells that can change do not have to be contiguous. We can enter the cells directly or we can point to the cells after clicking on the Expand button.

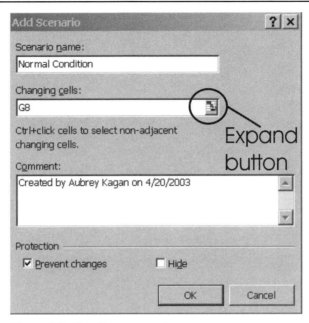

Figure 3-9: Creating a scenario.

First, we name the scenario. It is probably a good idea to start out with a normal condition and I have named it as such.

Clicking on the expand button on the right side of the **Changing Cells:** box will result in something similar to Figure 3-10. Clicking and **<Ctrl> + <Click>** will have the cells added to the box.

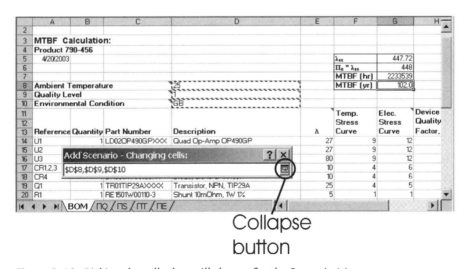

Figure 3-10: Picking the cells that will change for the Scenario Manager.

Clicking the collapse button on the right of the box will take us back to Figure 3-9. Click on **OK**. That will then take us to the next step where we must add the value that we want the scenario to evaluate. The dialog box in Figure 3-11 will cater for all the variable cells that we have defined.

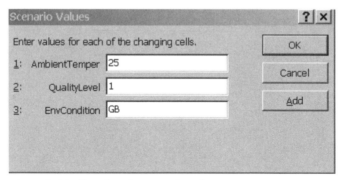

Figure 3-11: Entering the scenario values.

We enter the values that we would like to see evaluated, click on **OK** and our bidding is done. Following the same process (**Tools | Scenarios... | Add...**), we add scenarios, except that we no longer have to define which cells are going to change if we don't want to, as the cells we have already defined appear in the dialog box. See Figure 3-12.

We can then add the new values we want evaluated as in Figure 3-13.

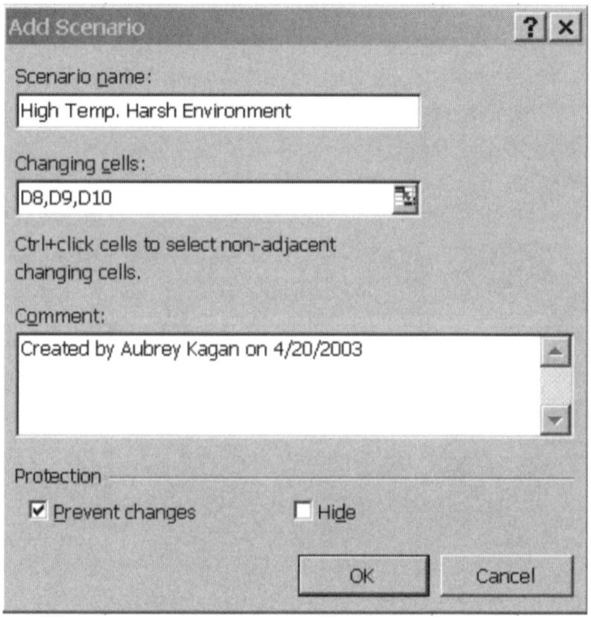

Figure 3-12: Defining an alternative scenario.

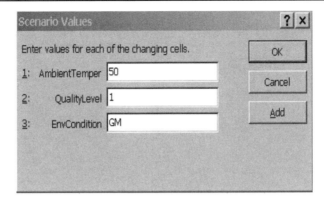

Figure 3-13: Values for new scenario.

Once we have completed all the scenarios, we can immediately view the results of a particular scenario by following the steps **Tools | Scenarios…**, click on the named scenario, and then **Show**. On simple examples, where the results can be condensed into a small area of the screen, it is possible to place the Scenario Manager dialog box so that selecting different scenarios shows the changes without having to close the dialog box as in Figure 3-14. This project is called *mtbf_scenario.xls* on the accompanying CD-ROM.

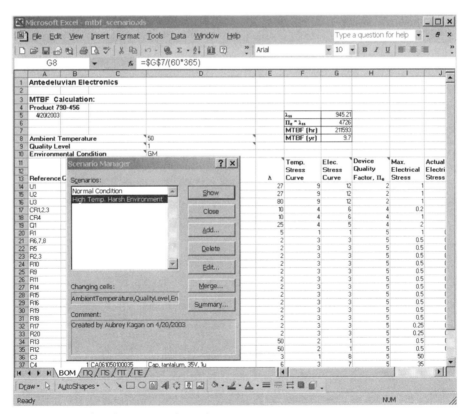

Figure 3-14: Choosing a scenario to view.

EXAMPLE 4

Counting Machine Cycles

Model Description

Despite the plethora of 16- and 32-bit processors, 8-bit devices still dominate the embedded market. Sometimes, we have to write very tight code and need to know the execution time by counting machine cycles. Revisions in the code mean reworking the machine cycle count. Looking up the number of machine cycles for each instruction is both labor intensive and prone to error. Cue the theme to the "Lone Ranger." Enter Excel to the rescue. The secret is Excel's ability to load almost any text file and interpret spaces and other characters as delimiters. Once the listing is loaded, we need to identify the cells with machine code and extract and manipulate the op-code to produce a numerical lookup value. Using a lookup table derived from the microcomputer's op-codes, this value is used to collect the associated number of machine cycles.

Each processor and compiler/assembler combination results in a different format of code so it is hard to provide a generalized model, but the principle remains the same. We will need to look at the idiosyncrasies of the compiler or the assembler in each case in order to format the model. Working with RISC type processors like the PIC is relatively simple in that all instructions take one machine cycle except jumps, which take two. Other processors, like the 8051, have different lengths of machine cycles for instructions, but irrespective of the conditional program execution, the time taken is the same. Yet more complex processors like the Z80 are a combination of both.

Since I have examples of 8051 code written in C, I will use this as a demonstration, but I will allow for the possibility that conditional instructions may take longer, just so you can see the approach.

Importing the File

The file I will be importing is a listing file produced from the source code written in C using the IAR Embedded Workbench Compiler (version 5.20).

In Excel, open the listing file in the usual way: **File | Open**, and then add *.lst in the File Name box and select the file that you want as shown in Figure 4-1.

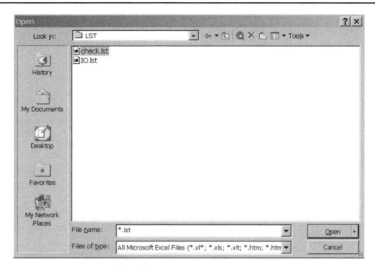

Figure 4-1: Open a list file.

Open the file and you will be presented with the Text Import Wizard, as shown in Figure 4-2.

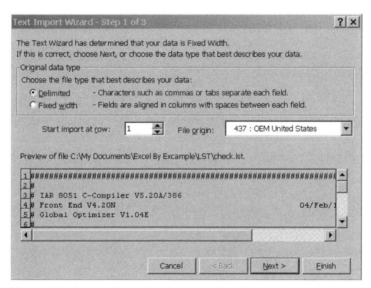

Figure 4-2: Step 1 of the Text Import Wizard.

Click on the **Delimited** data type. It is possible to remove a number of lines from the top of the file by starting the import at a particular row. Identify which row by scrolling down the preview window at the bottom of the screen, but you must manually enter the line number in the **Start import at row** box. Then click on the **Next** button. The next step in the File Import process is seen in Figure 4-3.

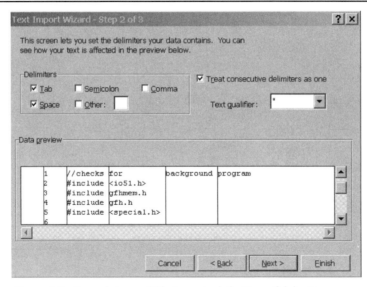

Figure 4-3: Second stage of file import, definition of delimiters.

Choose **Tab** and **Space** as delimiters and **Treat consecutive delimiters as one**. The Data Preview provides an idea of how the worksheet is going to appear. Select the **Next** button and the third stage (Figure 4-4) will be seen.

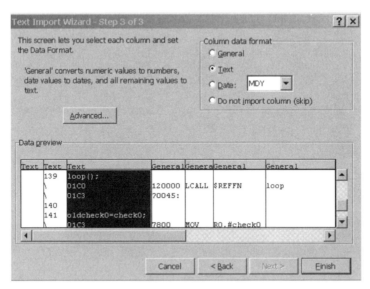

Figure 4-4: Third stage of file insert, formatting columns.

For each of the first four columns, click on the bar at the top and format the column as text using the radio buttons in the **Column data format** area. This is important because without this Excel will look at each cell in the op-codes and interpret as a number (when there is no alpha character) and text when there is. Classifying it as text will allow us to analyze the op-codes in a consistent manner.

Click on Finish to see the initial worksheet (Figure 4-5).

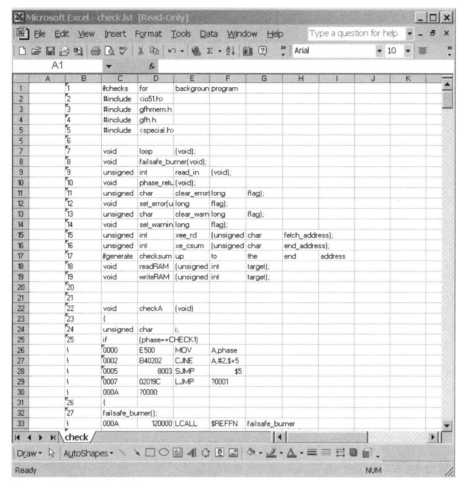

Figure 4-5: Initial worksheet.

Some comments in a line may stretch across many cells in a row, but this will not affect what we are trying to do. You will notice that the file importation results in only one worksheet.

Of particular note is that the compiler places a backslash "\" as the first character of any line with code in it, so I will use this as an indicator of a machine code entry.

Extracting Op-code

Rather than work on a large segment of code, for this example let's concentrate on the FOR loop that stretches from Excel line 110 to line 138 as shown in Figure 4-6. You will notice that I have done some additional preparation in the line in the form of headings for some of the columns (in bold).

In order to extract the op-code, we need to scan the cell in column B for each row in the range. If it is the backslash "\", then we must extract the first two characters of the string in the associated cell in column D. To do this, we will need the IF function followed by the LEFT function. We have seen that the IF function has the format

IF(*logical_test,value_if_true,value_if_false*)

For the logical test on row 110 we would enter =*IF(B111="\"*, ...

The format of the LEFT function is:

LEFT(*text,num_chars*)

In order to extract the first two characters in cell D110, this would become

LEFT(*D110,2*).

If there is no backslash, we want the column to be blank. If we put all that together as

=*if(B110="\", LEFT(D110,2),"")*

in cell I110, the result is a blank. Copying this cell to all the cells from I111 to I138 shows up all the op-codes as hexadecimal numbers, but formatted as text.

You will notice from cell I118 that the address label is also captured and since we would like to exclude it, we need to modify the formula. To this end, we note that the address text begins with the question mark. We can add a nested IF function so that the second term of the IF statement is the second IF statement:

IF (*LEFT(D110,1)="?","",LEFT(D110,2)*)

The complete statement would be

=*IF(B110="\",IF(LEFT(D110,1)="?","",LEFT(D110,2)),"")*

If we copy this to the cells in column I, the problem disappears. An alternative could have been to use the logical AND in the logical test, that is:

=*IF(AND(B110="\",LEFT(D110,1)<>"?"),LEFT(D110,2),"")*

Although Excel can handle text lookups, I chose to convert the text in column I to a number in order to use the lookup tables because the values in the lookup table are numeric. I have added this as another column in J to simplify the explanation. It is possible to combine the steps in the lookup function that we will use later. The entry in J110 is =*hex2dec(I110)* and this is copied down the cells in the range. Yet another hiccup shows up. Obviously from the results, applying this function to a blank cell returns a number 0. If we look this up in the op-codes it will return a valid instruction cycle time (NOP in the case of the 8051) and thus introduce an error into our calculations. We need to modify cell J110 *to =IF(I110="","",HEX2DEC(I110))*. In other words, if I110 is a blank, then so is J110.

This workbook is named *listing.xls*.

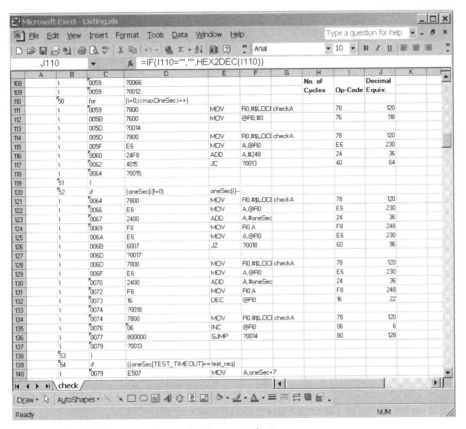

Figure 4-6: Initial extract of op-codes from a C listing.

Opening a Second Workbook

You will have noticed that there is only one sheet in this workbook. Inserting a file in this manner will always result in a single sheet. This proves troublesome in trying to generalize the model, since we would have to add a sheet and then copy and paste into the sheet. I suppose we could create a macro to do this, but it is not necessary. In Excel it is possible to access the contents of a second workbook as long as that workbook is open. We can create a workbook with all the op-codes and machine cycles as a separate entity especially since it does not need much in the way of modification once it has been created.

Let me describe the second workbook (8051opcodes.xls) as it appears in Figure 4-7. It can be opened while in any workbook using the **File | Open** sequence. Depending on the version of Excel, and probably the computer operating system, switching between the workbooks can be achieved by using the **Window** menu and selecting the workbook. Alternatively, you may be able to switch from the Windows taskbar as well.

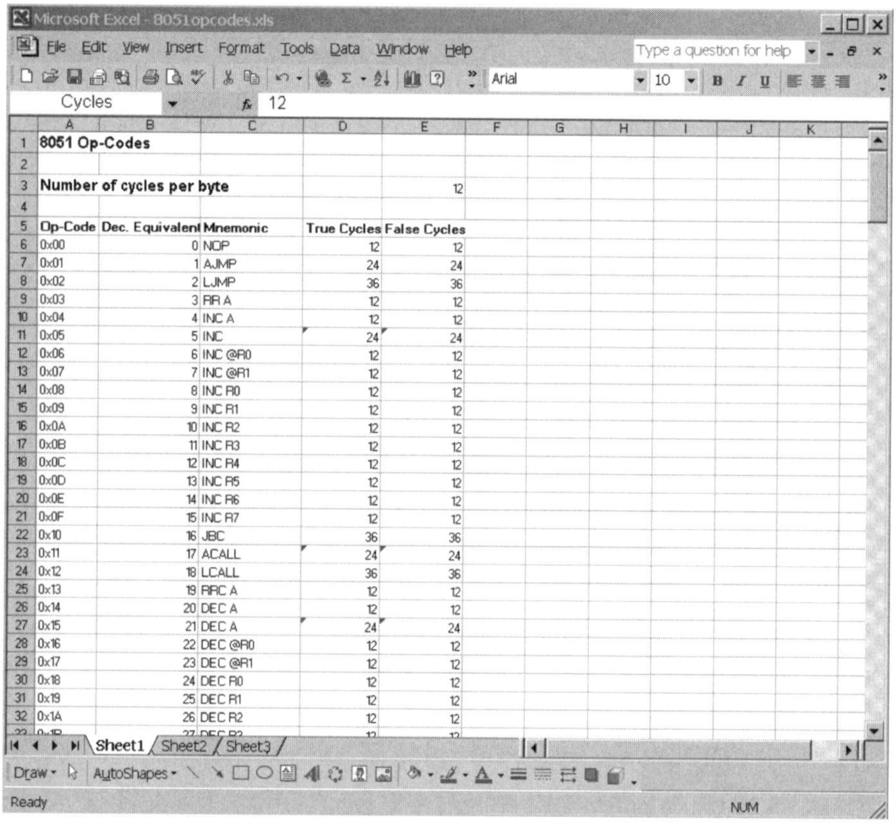

Figure 4-7: Lookup table of 8051 op-codes.

I "autofilled" from 0 to 255 in column B, and then created a hex format in column A. Converting the decimal number to hexadecimal requires a simple function in the form of:

DEC2HEX(number,places)

which will generate the hex number as a string to the length of the number of digits specified. In our case, in cell A6 this would be =*DEC2HEX(B6,2)*. This will generate a two-character string, but no indication that it is hexadecimal. It is easy enough to prefix the number with "0x" using concatenation. Excel has a special command for this:

CONCATENATE (text1,text2,…)

However, it also recognizes the ampersand as a concatenation instruction which is easier to use. Cell A6 becomes:

$$=(\text{``}0x\text{''} \& \; DEC2HEX(B6,2))$$

This cell is copied from A7 through to A261. Once this calculation is done, we really don't need the resources of the computer taken up by monitoring the value of cells that are not going to change. Select cells A7 through to A261 using **<Ctrl> + <C>** or one of the other alternatives to copy a block of cells to the clipboard. I have left A6 out so that the formula remains in the worksheet, in order to remember how this column was generated. With the same block still selected, click on **Edit | Paste special | Values** and the calculation is converted, nevermore to change.

Mnemonics have been added for aesthetic purposes, as well as to confirm the lookup while we check the operation of the worksheet.

I have reserved a cell, E3, for the number of cycles per byte read to allow for the newer versions of 8051s that execute in fewer machine cycles. Column D6 is a product of this cell and the number of bytes in the instruction. Although the 8051 executes the same number of cycles in a conditional statement, whether it is true or false, I have added a true and false column to illustrate how this may be used in the model for a different processor. If your processor does have different execution times, then this column will differ from the true column only in those specific instructions.

The range of cells B6 to E261 has been named *op_codes*.

In Parenthesis: *Recalculation and Auditing Formulas*

In this example, it may seem counter-intuitive that the value of cell A6 is derived from a value that appears later in cell B6. In early versions of spreadsheets, you had to consider the order that the spreadsheet evaluated cells (normally left to right and top to bottom), but today this is not normally an issue. A formula or cell is updated when any input that affects it is updated. **Tools | Options | Calculation** tab can affect how this calculation is done.

In some worksheets, the recalculation may take some time and can be inhibited by selecting the Manual button. Pressing the F9 key will force the whole workbook to be recalculated including custom functions, irrespective of whether the manual option is selected.

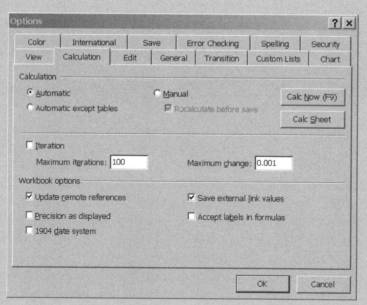

Figure 4-8.

It is possible to audit cells in order to track how one cell is affected by another, or even help to diagnose a problem. Follow the sequence **Tools | Formula Auditing**. This shows a bunch of options, but these are also available on a floating toolbar so click on **Show Formula Auditing Toolbar**, which will result in the following toolbar:

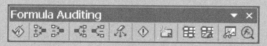

Figure 4-9.

Using the toolbar, it is possible to visually trace dependents and precedents of a particular cell (if indeed it does have connections). Each relationship has a tracer arrow marked as each cell is considered. The arrow is not removed until they are explicitly deleted individually or as a group. You can identify which button to use for a function from the pop-up tool tips.

When a formula results in an error, like #REF, click on the problem cell and use the **Trace Error** button to help analyze the problem. Clicking on **Circle Invalid Data** will highlight cells with invalid data in red to help with the debugging.

This is an invaluable tool in larger workbooks.

Cross Workbook Reference

At the risk of repeating myself, the vertical lookup function has the format:

VLOOKUP(*lookup_value,table_array,col_index_num,range_lookup*)

So in cell K111 of the listing.xls worksheet we enter:

=VLOOKUP($J111,'8051opcodes.xls'!op_codes,3,FALSE)

and we are rewarded when the number of machine cycles for the true condition appears. There is a problem in that there are cells in column J that have no value, so we need to embed the above formula in an IF statement as follows:

=IF($J111<>"",VLOOKUP($J111,'8051opcodes.xls'!op_codes,3,FALSE),"")

so that if a cell in column J is blank, the cell in K will also be blank.

In order to look up the number of execution cycles for a false condition, the formula in K111 is copied to L111 (hence, the absolute reference to column J). You will notice that I have added a column marked "Notes". In this column I have either a blank value, or the number 1 or 2. A blank value means that in a program with a loop, such as this example, this line is only counted once. The note "1" means that the instruction is executed once in each loop execution, and "2" is the conditional statement where it is executed once for each loop except for the final loop where it executes a (possibly) different number of cycles. In order to do this we use nested IF statements:

=IF(M111="",K111,(IF(M111=1,($H$110*K111),((($H$110-1)*K111)+L111))))

where $H$110 is the cell containing the number of cycles.

There is a bit more I have to say about the complexity of this nested IF, but I would like to leave this as a file on the CD-ROM so let me complete it first. We just need to sum all the entries in column N by blocking from N111 to N138 and clicking on the quick sum button (the Σ on the toolbar). This is the total number of cycles. If we divide it by the clock frequency (12 MHz in this case), we will have the execution time in seconds. This file is titled *listing.xls*, and the result is shown in Figure 4-10.

Easing the Pain of Nested IFs

As you can see from the formula, it can be difficult to keep track of the parentheses and the IFs. Imagine if you wanted to nest even more conditions!

Now is a good time to introduce the CHOOSE function. It has the format: CHOOSE(*index_num,value1,value2,…*)

Based on the index_num value, contents of the cell associated with that index are returned. In cell N111 the formula will be:

= CHOOSE(M111,K111,$H$110*K111,(($H$110-1)*K111)+L111)

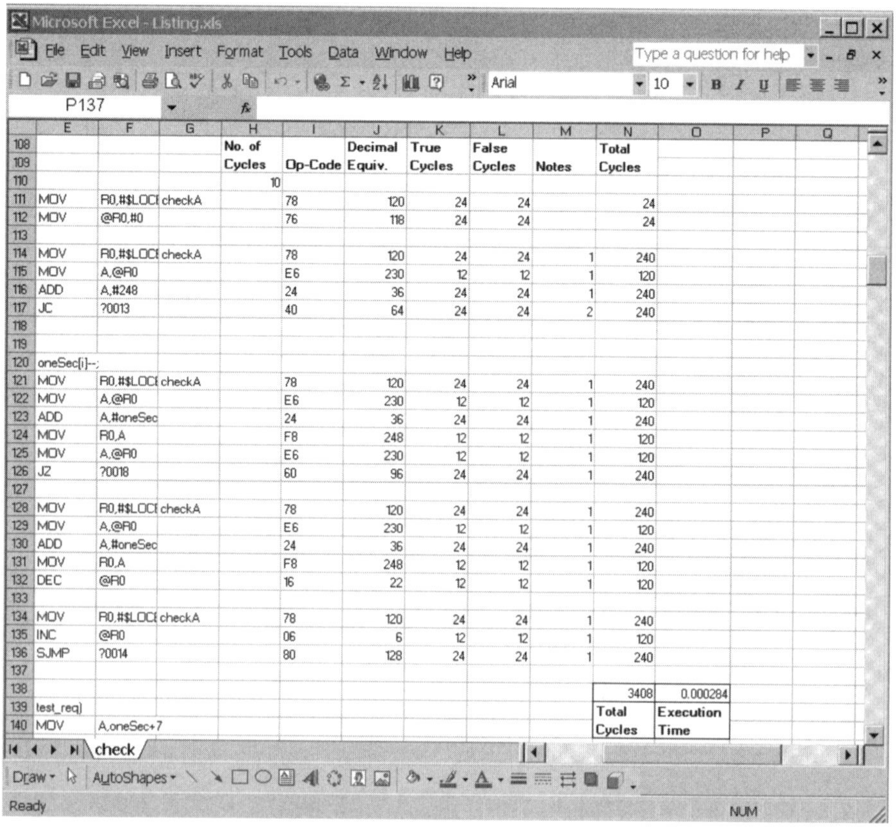

Figure 4-10: Completed calculation of execution time.

but if this formula was to occur on a blank cell in column M, an error would be generated. We obviously need to modify the formula with an IF statement and it becomes:

$$=IF(M111="","",CHOOSE(M111,K111,\$H\$110*K111,((\$H\$110-1)*K111)+L111))$$

This formula is copied through the range, and the project is complete and saved as *listing1.xls* on the CD-ROM. The appearance, however, is no different to Figure 4-10.

Unfortunately, since opening the list file results in a new worksheet, this method does not lend itself to automation. It is possible to cut and paste from the initial development into the new file, but you need to keep track of the size of the block that is copied and the size it is being pasted to. An alternative is possible by changing a line of the formulas to text and copying them to some other location, perhaps in 8051opcodes.xls. This is easy enough. To convert from a formula to text, either remove the = from the start of the line or add an apostrophe ' to the start of the line.

EXAMPLE 5

Character Generator

Model Description

There are many instances of dot matrix displays in use in electronics. Some devices, like the ubiquitous LCD modules that are based on the Hitachi standard, have a built in character generator. However, the Hitachi designers recognized that the user may want to have special characters and allows the creation of customized patterns. Other devices like the NKK Smartswitch (36×24 LCD display) or the Fairchild GMA2875 (5×7 LED display) require a character generator.

Every pattern or shape in a dot matrix display is defined by a series of active and inactive bits that determine if a pixel is on or off. Figure 5-1 indicates the formation of one such character.

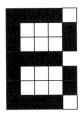

Figure 5-1:
5×7 dot matrix pattern for the letter "B".

Most dot matrix displays employ some form of multiplexing to drive the pixels. Depending on the display and the technique employed, the pixels can be generated for the horizontal axis or the vertical axis. If we take the example in Figure 5-1, and allocate a byte to each horizontal row, assuming the least significant bit corresponds to the rightmost pixel, the value of the first byte would be 30 (0x1e), the second would be 17 (0x11), and so forth.

While I was developing a controller for the NKK Smartswitch, I serendipitously came across a brilliant design idea by Alberto Ricci Bitti in *EDN*. The idea was to lay out the matrix on a worksheet to visually create the character and use Excel to automatically generate the number associated with a particular bit pattern. The example you are reading is an extension

69

of the idea, hopefully providing versatility for a wide number of applications and especially insights into the use of Excel. The idea for the original extensions was also published as my Design Idea in *EDN*.

Creating the Basic Workbook

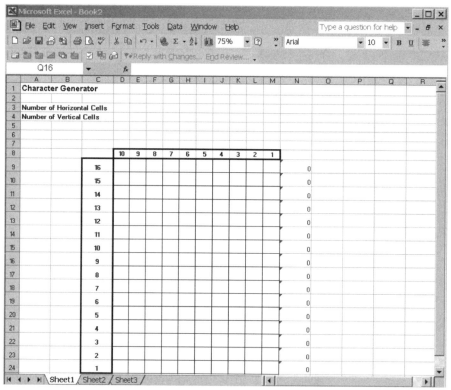

Figure 5-2: First setup of the workbook.

In order to generalize the idea, I allowed for a maximum size of the display graphic/character to be 10 columns by 16 rows. As is obvious from Figure 5-2, I intend to allow the user to enter the number of rows and columns so that the concept can be applied to smaller graphics. The matrix D9 to M24 represents the pixels on the display and I have formatted the width of the columns to approximate the height of the rows. By selecting the columns D to M (using the column select buttons) and right-clicking, it is possible to set them all to the same width. The same is true for the rows, although the value from row to column is not the same, so it requires a little trial and error. I also formatted the matrix for borders, bold text and centered alignment.

LEN Function

The crux of this project is the LEN function. It checks a string and returns the number of characters. In a row, we then check the contents of each of the columns from column D to M. If there are characters (nonzero LEN), then a number associated with this column is added to a running total. This is akin to the longhand process of converting a binary number to base 10. For example, 1011 binary is converted as $2^3 + 2^1 + 2^0$. Let's consider the Excel function:

=IF (LEN(D9),512,0)

Let's enter this in cell N9. If there is any character in cell D9, the value of cell D9 will be 512, otherwise it will be 0.

Now we need to consider all the cells in the row, so we change the entry in cell N9 to

=IF(LEN(D9),512,0)+IF(LEN(E9),256,0)+IF(LEN(F9),128,0)+IF(LEN(G9),64,0)+

IF(LEN(H9),32,0)+IF(LEN(I9),16,0)+IF(LEN(J9),8,0)+IF(LEN(K9),4,0)+

IF(LEN(L9),2,0)+IF(LEN(M9),1,0)

This is all entered as a single line without **<Enter>** or formatting. With this we have a simple method of generating a number for row 9. For the whole display, we copy this cell from D9 to D10 through to D24. Try clicking on a cell, entering a character, say **x**, and watch the numbers change in column D. We can now visibly set up the character we want and transfer the number for the rows to our software listing. Wait! There's lots more to come that can save you time.

Forms Controls

Some applications require the data to be shifted out horizontally and some vertically. It is a simple enough matter to transpose the above formula to generate the numbers for columns instead of rows. Instead of having both the horizontal and vertical calculations visible at the same time, we can add a drop-down control (called a *Combo box*) that will display the horizontal data when chosen, and blank it displaying the vertical selection when it is chosen.

Enter the data in cells N3, Q3 and Q4 as in Figure 5-3. N3 is a title, while Q3 and Q4 will be the entries in the drop-down box.

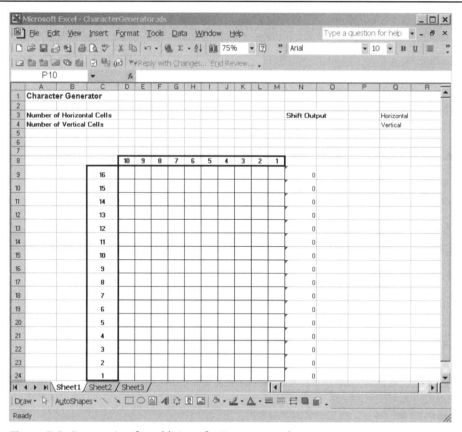

Figure 5-3: Preparation for addition of a Forms control.

In Parenthesis: *Forms Controls in a Different Version of Excel*

The use of Forms controls differs from one version of Excel to another. The Microsoft Knowledge Base has entries for the following versions:

Excel 2002: Q291073

Excel 2000: Q214262

Excel 97: Q142135

These are very handy in helping to understand the application of the controls.

We select **View | Toolbars | Forms** and then click on the symbol for the Combo box (note the pop-up tool help). Now we click around the top left corner of N4 and drag to form a rectangle. When you release the mouse button, you will see the Combo box as in Figure 5-4.

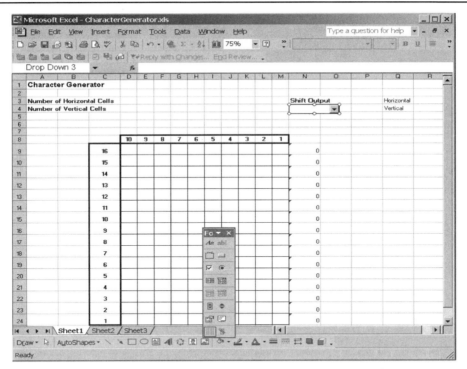

Figure 5-4: Placing a Combo box on a workbook. Note the toolbar at the lower center of the figure.

We have to link the entries to appear in the Combo box with the Combo box itself. As mentioned before, this is contained in cells Q3 and Q4. This information must be on the worksheet, although we can hide the column later. The resulting output of the Combo box must also be on the worksheet and it is a number associated with the Combo box selection. In our case with two possible options, it can be 1 or 2. We choose cell Q6 arbitrarily to contain the result. To connect the selection entries with the Combo box, we right-click on the Combo box and select **Format Control** and then select the **Control** tab. We select the input range, the cell link and the number of drop down lines. We terminate by clicking on **OK**. Now we click on any cell to take the focus away from the Combo box.

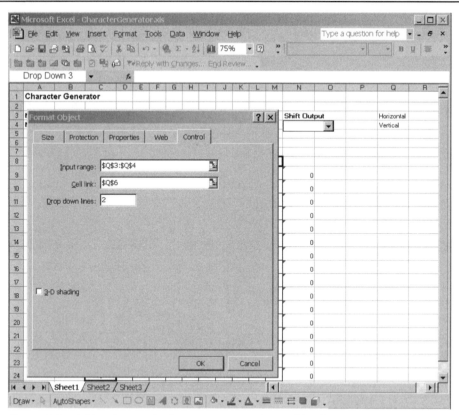

Figure 5-5: Linking the Combo box to entries on the workbook.

Clicking and making a selection in the Combo box will result in cell Q6 changing its value between 1 and 2.

We return to cell N9 in order to make a composite IF statement. The whole existing IF statement becomes the [value_if_true] part of the new IF statement. We need to insert the statement $IF(\$Q\$6=1,$ after the first "=" and add ,"") to the end of the statement, so that if the condition is not true, the cell becomes blank. The full line becomes:

$=IF(\$Q\$6=1,IF(LEN(D9),512,0)+IF(LEN(E9),256,0)+IF(LEN(F9),128,0)+$

$IF(LEN(G9),64,0)+IF(LEN(H9),32,0)+IF(LEN(I9),16,0)+$

$IF(LEN(J9),8,0)+IF(LEN(K9),4,0)+$

$IF(LEN(L9),2,0)+IF(LEN(M9),1,0),"")$

We now copy this statement to cells N10 through to N24, and hide column Q. Changing the Combo box selection will toggle the cells N9 to N24 from blank to a number. To get the vertical shift ability, cell D25 must contain a formula that will allow for 16 entries. It should be:

=IF($Q$6=2,IF(LEN(D9),32768,0)+IF(LEN(D10),16384,0)+IF(LEN(D11),8192,0)+

IF(LEN(D12),4096,0)+IF(LEN(D13),2048,0)+IF(LEN(D14),1024,0)+

IF(LEN(D15),512,0)+IF(LEN(D16),256,0)+IF(LEN(D17),128,0)+

IF(LEN(D18),64,0)+IF(LEN(D19),32,0)+IF(LEN(D20),16,0)+IF(LEN(D21),8,0)+

IF(LEN(D22),4,0)+IF(LEN(D23),2,0)+IF(LEN(D24),1,0),"")

The formula should be copied into cells E25 to M25.

Text Orientation

When the number in a column contains a large number, the cell shows "###" to indicate that the column is too narrow to display the result. This would be inelegant in the cells associated with the vertical shift. There is an easy way around it by formatting the cells from D25 to M25 to an orientation of 90 degrees, and then sizing the height of row 25 accordingly.

Comments

Some of us may have noticed a green triangle in the left-hand corner of some cells. This is an indicator that Excel considers the result of a calculation that is contained in the cell to be potentially suspect. See "In Parenthesis: Excel Warning Detection" in the next example for a more detailed explanation.

There can also be a red triangle in the right-hand corner of a cell, which indicates a comment. As in programming, a comment is placed by a programmer in order to help understand and remember the reasons for a particular approach. Excel's comments though, provide a pop-up window when the cursor hovers over the cell. This feature can be used to guide a user or the programmer as to what number to enter or any other useful information.

Adding, editing and deleting a comment is no more difficult than right-clicking on a cell and choosing the relevant menu entry. The entry box is a mini-word processor and we can paste and cut text, size the window and more.

I have added a comment to cell N3 to describe the use of the function associated with this cell. I will be adding other comments to the workbook as we go although I will not mention them. When starting, the comment box is initialized with the registered user's name. This is ordinary text and can be edited as you like. I just delete it. Comments are hinted at by the red triangle, and can all be viewed simultaneously using the menu sequence **View** | **Comments**. Repeating the sequence will return to the pop-up mode.

Double-Click Macro

Moving around the workbook, typing a character and moving with the **<Enter>** or arrow keys is inefficient. A macro that uses the double-click of the mouse to toggle between a character and no character would make the application much more elegant. And while we are at it, we will add a black shading (that actually hides the character) and gives a better approximation of the appearance of the pattern.

Open the Visual Basic Editor and right-click on **Sheet1** under Microsoft Excel Projects. Select **View Code**. In the left-hand drop-down box select **Worksheet**, and in the right-hand box select **BeforeDoubleClick**.

Enter the following code:

```
Private Sub Worksheet_BeforeDoubleClick(ByVal Target As Range, Cancel As Boolean)
    'checking if we add or remove data
    If Target = "x" Then
       Target = ""
       'blanking cell entry
       Selection.Interior.ColorIndex = xlNone
       'using no shading colour

    Else
       Target = "x"
       'setting the cell contents to a string of finite length
       'so that the number generator sees it.

       'and then turning the shading black to obscure the x

       With Selection.Interior
          .ColorIndex = 1
          .Pattern = xlSolid
          .PatternColorIndex = xlAutomatic
       End With

    End If
End Sub
```

Because this procedure runs on an event occurrence (double-clicking on a cell), there is no need to explicitly invoke it.

Pretty easy stuff, but the result is great. Only one drawback, once we have toggled a pixel on or off, we need to change the cell focus (click in another cell) before returning to the original cell to toggle it again. Actually, there is another failing. It is possible to double-click in any cell and this character and shading effect will happen. We can mostly get around that by protecting cells (or by testing for valid cells in the double-click procedure).

In Parenthesis: *Cell Protection*

Changing cells containing formulas can cause some problems for the workbook. The simplest solution is to protect the cells so that it is impossible to change. Excel implements protection in two stages. Initially, we set the cells we want to unprotect (Excel assumes all cells are protected unless informed otherwise). This does not initiate the protection yet. In the second stage, the worksheet is protected and now the protected cells are inviolate.

In order to achieve cell protection, select the cells that are to be unprotected and then right-click and choose **Format Cells** and the **Protection** tab. Ensure that the **Locked Cells** is unchecked and click on **OK**. Then follow the sequence **Tools | Protection | Protect Sheet**. Check and uncheck the boxes that are pertinent. There is no need for a password if you don't want one. Click on **OK**. The user can now modify the unprotected cells.

In the **Format Cells** dialog there was a **Hidden** checkbox. If it is selected and you subsequently click on a cell with a formula with the worksheet protected, the formula will not be displayed in the formula bar. Note that one of the options when protecting the sheet is to disallow selection of protected cells, so the user can never get there if we so wish.

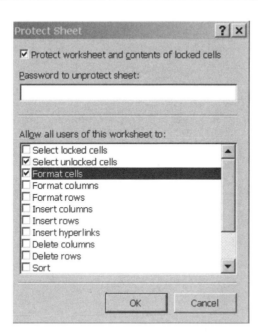

Figure 5-6: Options on protecting a sheet.

Disable protection on cell block D9 to M24. Unhide column Q and disable protection on cell Q6. Unhiding can only be done when the sheet is unprotected. If you are reading this paragraph in sequence, then the sheet is still unprotected. If you are rereading it, trying to get the Shift Output drop-down to work, then **Tools | Protection | Unprotect Sheet....**

Since the **Select locked cells** is unchecked, the user will be unable to click on any of the locked cells. We must check the **Format cells** option or the Double-click macro will not be able to change the shading on a cell.

Once we are satisfied on the operation, we must unprotect the sheet in order to allow some of the new features we are going to add. We must just remember to reprotect at the completion of the project. The individual protection remains with each cell.

Macro Activation by the Command Button

So far, the model requires users to determine their limits on the screen when working with matrix less than the maximum size. We are going to add a macro that will take the contents of cells D3 and D4 and shade the matrix to provide an indication of the nonusable area. This macro will be run from a pushbutton on the workbook.

First we must ensure that cells D3 and D4 are unlocked. We will also format the cells as bold for cosmetic reasons.

Create a macro in the Workbook area as follows: **Tools | Macro | Visual Basic Editor**, right-click on the **This Workbook** "folder" and select View Code. Select **(General)** from the drop-down box on the left of the code window. In the code window, enter

Sub ShadeMatrix_Click followed by the <**Enter**> and Visual Basic will automatically initiate the routine.

> **In Parenthesis:** *Cells Notation Versus String Manipulation*
>
> The code used in this example accesses cells on the worksheet and works by creating strings. Microsoft considers that this is not good programming practice as detailed in the VBA Help subject "Range Collection". It is much easier to use the alternate notation CELL(row,column) to read or write to a cell.
>
> Consider me admonished.

Enter the macro as follows:

```
Sub ShadeMatrix_Click()
'
' shade Macro
'

    Dim sX As String
    Dim iUtil As Integer

    ' clear all the shading
    Range("D9:M24").Select
    Selection.Interior.ColorIndex = xlNone

    'and clear all the contents
```

Selection.ClearContents

If Range("c9").Value > Range("D4").Value Then
'to allow full range of operation

sX = "D9:M"
'initiate the horizontal range as a string
iUtil = (Range("c9").Value - (Range("D4").Value + 1)) + 9
'by using the value in C9, it is possible to create larger
'matrices without changing the macro
'D$ has the number of rows
sX = sX & iUtil
'this easy concantenation merges a string and a number into a string
'sX is has a range now stretching down from row 9
'Note how easy string manipulation is.
Range(sX).Select
'shading the rows
With Selection.Interior
   .ColorIndex = 16
   .PatternColorIndex = xlAutomatic
End With
End If

If Range("d8").Value > Range("D3").Value Then
'to allow extremes of range
'create the vertical range
iUtil = 68 'D in ascii
'since we have to figurue out how many columns
'we need to work arithmetically with the column identifier (a letter)
iUtil = iUtil + ((Range("d8").Value - Range("D3").Value) - 1)
'by using the value in d8, it is possible to create larger
'matrices without changing the macro
sX = Chr$(iUtil)
'converting the calaculated value back to text to use
'as a column identifier
sX = "D9:" & Chr$(iUtil) & "24"
'concantenating strings
Range(sX).Select
'once the area is defined we select it
With Selection.Interior
   .ColorIndex = 16
   .PatternColorIndex = xlAutomatic
   'and shade it
End With

End If

'place pointer at top left hand of allowed frame

IPoint = 24
IPoint = IPoint - (Range("D4").Value - 1)
sX = IPoint
'type conversion
sX = "R" & sX & "C"
'concantenate for rows and add C for columns
IPoint = 13
IPoint = IPoint - (Range("D3").Value - 1)
sX = sX & IPoint
'should do the conversion as well
Application.Goto Reference:=sX

End Sub

Sequence through **View | Toolbars | Forms** and click on the **Button** icon. Click in the top left-hand corner of cell G3 and drag until the button is a suitable size. When we release the mouse button, a dialog to name the macro pops up as Figure 5-7. Select the macro we have just entered, **ShadeMatrix_Click**, and **OK**.

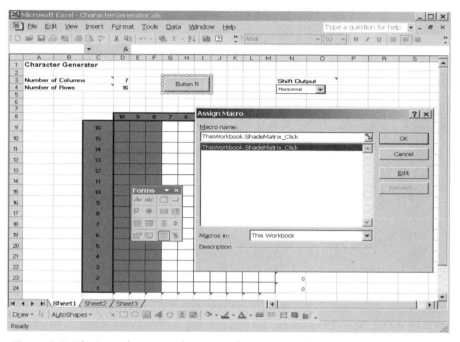

Figure 5-7: Placing a button and naming the associated macro that runs when the button is clicked.

If not already selected, select the button by right-clicking. Then click on the button and edit the caption to read "Shade Matrix" and click away from the button.

Save to Data File

All this is well and good, but in creating a whole character set it is a good idea to reduce keystrokes. Rather than cut and paste each character to the text editor of the compiler, we are going to create a macro that produces a text file (if it does not already exist) and then adds a character definition to that file every time the macro is run. If we create a command button to run this macro next to the matrix, the process becomes very simple since it is all mouse controlled. When the button is clicked the user is prompted for a comment about the character, so we could simply enter the character associated with the pattern, or something more complex. Once the character is stored, the "canvas" is cleared in readiness for the next character. This target file is intended to be opened with a text editor and the contents copied to the source file. Since there are many assemblers and variations of C (and even other languages) we should allow for maximum versatility.

We create the cells A26 to B29 as in Figure 5-8. Unlock cells B26 to B29.

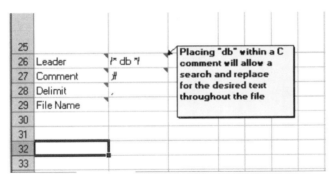

Figure 5-8: User input to allow for customization of text file output.

The leader is placed at the start of a line in order to define a series of constants. In some assemblers this is a "db" directive. I could foresee that there may be some time when in one project I would need the assembler character set and at a later stage, the same character set in C. If we use the "/* db */" format then, using the editing capabilities of the word processor it would be possible to search and replace as necessary. The comment field is the string that will be used as the prefix to cause the compiler/assembler to ignore the comment.

The delimit entry is the character that is used to separate the bytes in the source code. If the delimiter is not accessible from the keyboard (like the **Tab** code), we can enter the formula:

=CHAR(xx)

where xx is the numerical representation of the character. For example, the space character would be 32.

Finally, the File Name is the name the the data will be stored in with the suffix ".txt". In addition, it will be used as the first line of the file appended to the comment character above.

Now we create a macro as follows:

```
Sub SaveCharacter_Click()
'
' SaveCharacter Macro
' Macro recorded 11/16/01 by Aubrey Kagan
'
' Keyboard Shortcut: <Ctrl>+<Shift>+<T>
'
'if file does not exist then create it.
    Dim sFname As String
    Dim sX As String
    Dim iPoint As Integer
    Dim sComment As String
    Dim iUtil As Integer
    Dim iFileExists As Boolean
    Dim sX2 As String

    sFname = Range("B29").Value
    'fetch the file name and tag the suffix on
    sFname = sFname & ".txt"

    sX = Dir(sFname)
    'if the file does not already exist, then create it
    If sX <> "" Then
        iFileExists = True
    Else
        iFileExists = False
        Open sFname For Output As #1

        'take the comment symbols an place the file name as the first
        'comment
        sX = Range("B27").Value & sFname
        Print #1, sX
        Close #1
        'file created and saved, with the first line
    End If
    'now we print a range of values across the page
    'associated with the bytes for the pixels
    If (Range("Q6").Value) = 1 Then
    'for horizontal shift
        iPoint = 24
        'iPoint is the highest numerical value of row of the matrix
        iPoint = iPoint - (Range("D4").Value - 1)
        sX = iPoint
        'converting typ to string
        sX = "R" & sX
        sX = sX & "C14"
        'pointing at the start of the data
        Application.Goto Reference:=sX

        'get character description
        sComment = InputBox("Enter you comment for this Character", "Comment Creation")
```

```
      Open sFname For Append As #1
      Print #1, Range("B26").Value;

      For iUtil = 0 To (Range("D4").Value - 1)
        sX = (iPoint + iUtil)
        sX = "R" & sX
        sX = sX & "C14"
        'placing the cursor at the data address
        'column 14
        Application.Goto Reference:=sX
        Print #1, ActiveCell.Value;
        Print #1, Range("B28").Value;
        'print value followed by the delimiter
      Next
  Else
  'for vertical shift
      iPoint = 13
      'equal to column M
      'iPoint is the highest numerical value of column of the matrix
      iPoint = iPoint - (Range("D3").Value - 1)

      sX = "R25C" & iPoint
      'string type convesion and concatenation

      'pointing at the start of the data
      Application.Goto Reference:=sX

      'get character description
      sComment = InputBox("Enter you comment for this Character", "Comment Creation")

      Open sFname For Append As #1
      Print #1, Range("B26").Value;

      For iUtil = 0 To (Range("D3").Value - 1)
        sX = "R25C" & (iPoint + iUtil)
        'placing the cursor at the data address
        'row24
        Application.Goto Reference:=sX
        Print #1, ActiveCell.Value;
        Print #1, Range("B28").Value;
        'print value followed by the delimiter
      Next
  End If

  sComment = Range("B27").Value & sComment
  Print #1, sComment


  'ensure all files are closed
  Close
  'as a final step clear the page for the next character
  Range("D9:M24").Select
  Selection.ClearContents
  'move to the top left hand of screen
  'first deal with rows
```

```
iPoint = 24
'for row 24
iPoint = iPoint - (Range("D4").Value - 1)
sX = iPoint
'type conversion
sX2 = iPoint
'sX2 is used to clear the shading.

sX = "R" & sX & "C"
'concantenate for rows and add C for columns
iPoint = 13
'for column M (A=1, B=2...)
iPoint = iPoint - (Range("D3").Value - 1)
sX = sX & iPoint
sX2 = Chr(iPoint + (65 - 1)) & sX2
'creating the ASCII of the column
'and combining with the row for the cell location

'getting rid of the pixel shading

Range(sX2 & ":M24").Select
Selection.Interior.ColorIndex = xlNone

'and going to the top left cell
Application.Goto Reference:=sX


End Sub
```

Once this is done we return to the workbook. After creating a command button, starting in cell P12 (see Figure 5-9), we link it to the SaveCharacter_Click macro. The caption in the button should also be changed to something more relevant. We must remember to protect the worksheet and then save. *CharacterGenerator.xls* is the file of this model on the accompanying CD-ROM.

Usage

One thing to note: When changing the data in any cell and then needing to click on a button, until that cell is deselected, the button will not activate. We simply have to click anywhere on the worksheet, protected area or not, before clicking on a control button.

Depending on the zoom settings of the worksheet, Figure 5-9 is how the screen of the completed application would look. The instructions to use this application would be:

1. Enter the number of rows and columns in the display and click on the **Shade Matrix** button.

2. Select a horizontal or vertical shift from the drop-down box.

3. Enter the information required in cells B26 to B29. If the file name already exists, the information will be appended to the file.

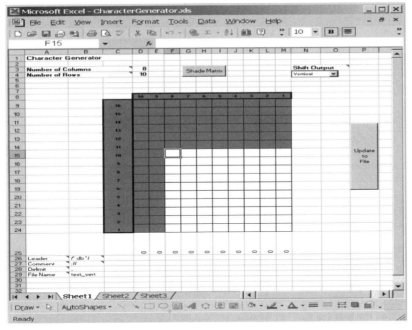

Figure 5-9: Completed application.

4. Create the pattern for the first character/image by double-clicking in a cell to activate a pixel and double-clicking a second time to deactivate the pixel.

5. When the character/image is complete, click on the **Update to File** button.

6. At the prompt in the message box, enter a comment for the help identify the character just entered, or any other text.

7. Every time this is done the tile is opened, the character data is added and the file is closed, so the application can be terminated at any time.

8. Repeat for all the characters in the set. Since the file is closed every time and because the application simply adds data to the existing file it is possible to create the character set over several sessions.

9. Edit the resultant file as necessary to match the computer language being used. It is also possible to edit the actual data if errors were made during entry.

10. Copy the file into a source file for the project.

Did you think Excel could do this kind of application?

EXAMPLE 6

8052 Microcomputer Register Setup

Model Description

Most modern microcomputers have a graphical user interface where the register configuration can be done through the modern paradigm of mouse clicks, graphical depiction and setup value calculation. Once you have worked with this approach, the return to the "good old days" for a more venerable processor may not be appreciated. Since you are reading a book on Excel, you could be forgiven for jumping to the conclusion that it is possible to do this in Excel. Of course you would be right!

This is a fairly simple, if somewhat lengthy application, but it gives me a great opportunity to showcase Forms controls while demonstrating several other features of Excel.

I still remember the feeling of freedom migrating from the Intel 8048 microcomputer to the 8051. Although by today's standards the resources of the 8051 are limited, obviously the Intel designers covered a large segment of the market as proven by the longevity of the product. The structure of the peripherals in the device is controlled by configuring a series of registers. I have broken the register structure into four sections: the input/output, the counter/timers, the serial communications port and the interrupt structure. I have further broken the counter/timers into three sub-groups. These sheets will be named: *I_O, C_T0, C_T1, C_T2, Serial* and *Interrupt*.

Spreadsheet Concept

For each section, I created a worksheet in the workbook named *8052.xls*. I added and named sheets according to the microcomputer hardware partition as listed above. This can be done by right-clicking on the sheet tab, and selecting **Insert** or **Rename**. It is also possible to rename by double-clicking on the Sheet tab.

To place controls we need the Forms toolbox, which we invoke from **View | Toolbars | Forms**. Starting with sheet I_O, we place a Group box in the upper left-hand corner by clicking on the Group box button and then clicking in the worksheet and dragging to a suitable size. In order to change the Group box name, it must first be selected and then you can click on the text, or right-click and select **Edit Text**. I renamed it to *Processor Selection* since I am going to select between the 8051 and 8052. It is prudent to place the Group box before placing the controls that it contains, or the controls may not be associated. The size and placement of the Group box can be changed when the box is selected. The size is changed by the "handles" on the selection window when the cursor changes to a two-headed arrow. The position can be changed by dragging the Group box when the cursor changes to a four-headed arrow over the frame of the box.

In Parenthesis: *Forms Control*

The Forms control should not be confused with the Control Toolbox, which is for inserting ActiveX controls. The Forms controls are easier to understand and use than ActiveX controls, but they are less flexible. To get to the Forms toolbox in Excel 2002, follow the sequence **View | Toolbars | Forms**. The toolbar that appears is similar to this:

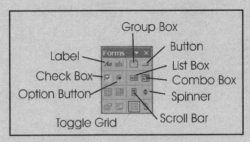

Figure 6-1.

Every Excel worksheet has an invisible layer. Like graphs and graphics, the tools are stored on this layer and appear to "hover" above the worksheet. It is possible to have an Excel entry behind it.

We have all experienced these controls in many Windows applications. The Label tool allows text to be entered and can be used to provide a title to a control. The Check box allows the user to select or remove a feature. The Option buttons are similar, but are linked so that only one feature of the group is selected at a time. The Group box collects objects like the Option buttons into a logical group. If you have two or more groups of Option buttons, then they must be placed in two Group boxes or Excel will only allow one or all of them to be selected. The Button control, when suitably configured, will run a macro when it is "pressed."

The List box shows a list of a number of options (with a Scroll bar when the number of options exceeds the box size). The user clicks on one to make a selection. The Combo box is similar, but only displays the list when the drop-down control is clicked. Once selected, the selected option is shown in the box.

A Scroll bar allows the user to adjust a value from a maximum to a minimum and back. It allows for three vernier levels. Coarse adjustment is by dragging the Scroll bar; medium by clicking within the gap between the Scroll bar and the direction arrows; and fine by clicking on the direction arrows. The result appears in an Excel cell and it is associative—i.e., if a value is entered in the cell, the Scroll bar will move to the appropriate position.

The Spinner is a set of up and down arrows that will increase or decrease a value.

Although not really a control, the Toggle Grid makes the background grid of lines visible or invisible.

I placed two Option buttons in the Group box by clicking on the Option button icon in the Forms toolbar, then clicking within the "Processor Selection" Group box and dragging to a suitable size. Placement of the buttons is a hit and miss affair since it appears that Excel does not provide tools to align them, but does provide tools to size them, although not simultaneously. (Within Visual Basic however, these restrictions do not exist though this does not help us here.) I renamed the Option buttons *8051* and *8052* since the difference between the NMOS and the CMOS versions is only in the power saving modes, and it does not form part of the initial setup of the device.

Once placed, the control becomes functional so that clicking on it will not allow us to change the properties. We can only modify the properties by right-clicking on the control.

We have to decide on a cell that will carry the resultant of this group of Option buttons. Depending on which button is selected, this cell will have a value representative of the button. Right-click on either of the Option buttons, and select **Format Controls** and then the **Control** tab. We direct the Cell link to cell **A6** (see Figure 6-2). Since these Option buttons are linked, we only need to set up this cell link for one of them. Now click away from the button. Click on **8051** and check that cell A6 has the value of one. Click on **8052**, and notice that the value of A6 changes to 2. Since A6 will be used in later processes, this value must always be there, but at a later stage we will hide the column.

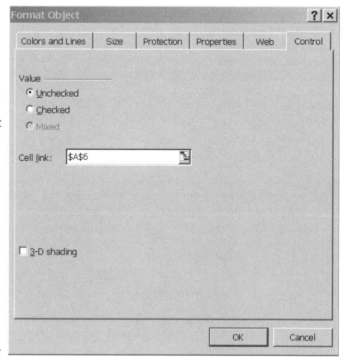

Figure 6-2: Formatting the Option button.

Throughout this example, I will also place the outputs (and inputs where they exist) in the rows below the Check box control so that they can be hidden, thereby improving the aesthetics.

In the **Format Control** for the Option buttons, there are other options that you can consider. Each button can have a line around it, the background color filled in, or even a 3D shading effect. I just opted for a line around, which actually helps with the alignment.

I also named cell A6 *Processor*.

Next we want to select the processor oscillator frequency, so I placed a horizontal Scroll bar and above it I placed a Text box. The completed result can be seen in Figure 6-4. The Label box format capability is rather thin, with far fewer format capabilities than text in a regular Excel cell. However, it does have one advantage in that it can be shifted around to just the right location. Of course there is font consistency between all the text entries in the controls. Figure 6-3 shows the control format capability of the Scroll bar.

Since the processor will run from 0 to 12 MHz, these limits determine the minimum and the maximum values of the Scroll bar. Clicking the left and right arrows is the "Incremental change" on the Scroll bars and we want this to increment by 1 each time this happens. Clicking in the gap between the slider bar and the incremental arrow is termed the "Page change" and we set this to 5. I have linked the changes to cell F4. Click away from the control, and

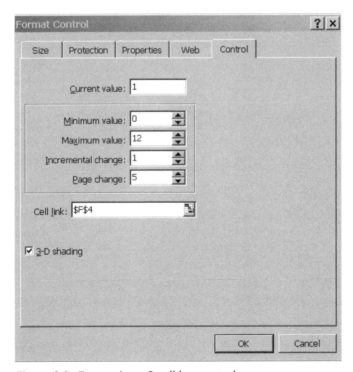

Figure 6-3: Formatting a Scroll bar control.

then move the slider bar around and note how cell F4 changes. Of course, it is conceivable that the crystal frequency will be a value other than in integer between 0 and 12. Simply clicking on cell F4 and changing the value to whatever the crystal frequency is (in MHz) will reverse the process and the slider bar will move to the appropriate place. I named cell F4 *Oscillator*.

To date, the worksheet appears as Figure 6-4.

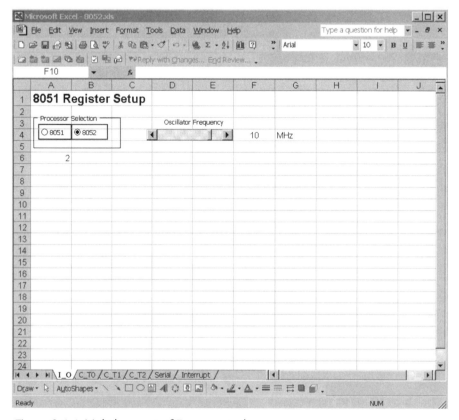

Figure 6-4: Initial placement of Forms controls.

From a configuration point of view, the port pins of the 8051 are simple. The only thing that needs to be set up is if a pin is an output, then it may need to be cleared to 0 on startup. We will use Check boxes to decide which inputs must be initialized to 0. After creating a table for the 8 bits of Port 0 in the worksheet range A8 to H10 (as in Figure 6-5), we have to place eight Check boxes. You will remember from the Option buttons how difficult it was to align just two. The description of one way to align the Check boxes follows. Now that this approach is in print, no doubt someone will let me know of a simpler way.

First, place a Check box anywhere on the worksheet. Delete the text and size it to as small as it will go. Click away from the box. Right-click on the Check box and select **Cut**. Then click in cell A11, right-click and **Paste**. Repeat for cells B11 through to H11. We should now have eight Check boxes, all offset to the same degree from the midpoint position of cells A11 to

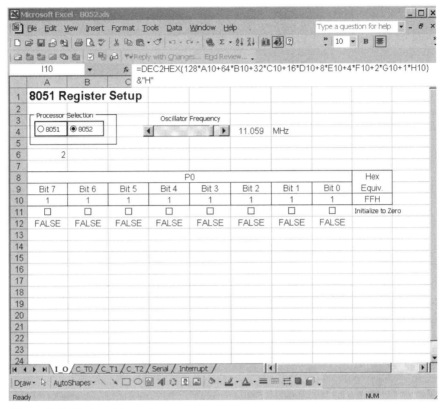

Figure 6-5: First register setup.

H11. Left-click on the one in column A. A menu may pop up, but ignore it and press **<Ctrl>** and then click on the remaining seven Check boxes allowing a multiple selection. Move the cursor within this selection until the cursor changes to a four-headed arrow. Click and drag the boxes to the desired position. As before, each Check box requires a cell associated with it, and we do this with the cell below each box, for example, A12 through to H12.

Checking and unchecking each Check box will change the cell beneath it from FALSE to TRUE and back again. For bit 7, in cell A10 we enter:

$$=IF(A12=TRUE,0,1)$$

So it shows a 1 if the box is unchecked and a 0 if it is. We copy this formula to all the bits in cells B10 through to H10.

At the right-end of the table, we will add the hexadecimal byte associated with this setup. Cell I10 contains the formula:

$$=DEC2HEX(128*A10+64*B10+32*C10+16*D10+8*E10+4*F10+2*G10+1*H10,2)\ \&"H"$$

Within the parenthesis is the numerical conversion based on the binary value of each bit. The DEC2HEX function converts the number to hexadecimal (with 2 characters derived from the last parameter in the junction), and then the "&" concatenates the "H" suffix. The overall appearance will look even better when we hide row 12.

Next, we block from A8 to I12 and copy to locations A14, A20 and A26. In this manner, the Check boxes are copied as well. Note that when "showing" the cell using the expand button, an absolute address is inserted. When you copy this, if this remains as absolute we will need to edit every control after the copy. It is probably a good idea to remove the $ signs. I should note that doing this can be quite frustrating, since the arrow keys don't move the cursor but add in cell information. You will need a combination of mouse clicks and **Backspace** or **Delete** keys.

We then edit the title of each table to the correct port ID. Later in the process, we will want to protect the whole workbook to prevent the user from changing the controls and formula. To allow this to happen we must first unlock the following cells: F4, A6, A12:H12, A18:H18, A24:H24 and A30:H30. The process is as follows: select the cell(s), right-click, **Format Cells | Protection** and make sure **Locked** and **Hidden** are unchecked.

After hiding rows 6, 12, 18 and 24, we click on **Toggle Grid** on the Forms Control toolbar to see Figure 6-6.

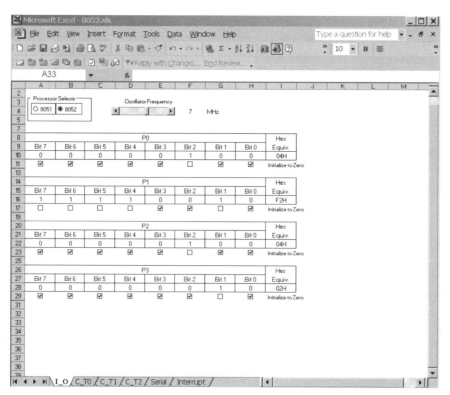

Figure 6-6: First sheet showing processor selection, oscillator frequency and port initialization. Note the lack of the grid.

We will improve further on this appearance by the end of the model development, but for the moment, I suggest clicking the **Toggle Grid** again so we can develop further.

Counter/Timer 0 Sheet

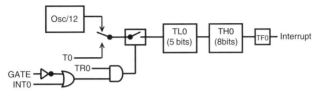

Figure 6-7: Timer/Counter 0, Mode 0.

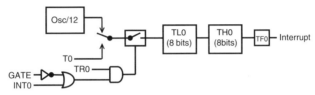

Figure 6-8: Timer/Counter 0, Mode 1.

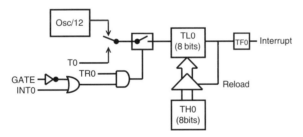

Figure 6-9: Timer/Counter 0, Mode 2.

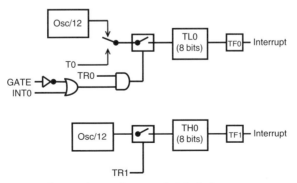

Figure 6-10: Timer/Counter 0, Mode 3.

Let's move along to the next worksheet marked "C_T0". Counter Timer 0 can be operated in one of four modes (see Figure 6-7 to Figure 6-10), and to select the mode we will implement a Combo box control. The options that drop down in the Combo box selection must appear in the workbook. One of the shortcomings of this Forms Control is that the list must be vertical, it cannot be horizontal. (Another shortcoming is that it is not possible to make the controls visible or invisible, unlike ActiveX controls where it is possible with similar controls. The limitation has influenced my approach to this model.) As with the other controls that we have used, there also needs to be a cell link to hold the output. As you can see in Figure 6-11, the text input for the options is in cells A5 to A8, and the linked cell is A4. Cell A4 has also been named *TC0_Mode* in a separate step and is unprotected.

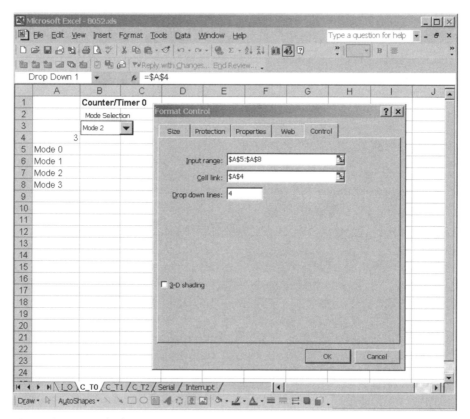

Figure 6-11: Combo box setup.

The timer/counter can be configured as a timer or a counter. The essential difference is the source of the clock for the up counter, as can be seen from the switch between OSC/12 and T0 in Figure 6-7 to Figure 6-10. In principle there are only these two options, but conceptually, there is an added dimension if we want to add some sophistication to the application.

Let's consider if we want to measure a pulse width using Mode 0 (Figure 6-7). We would want the input to be on INT0, and the switch C/T set to the C option. When the signal on

INT0 is high (with the configuration bits GATE and TR0 both set to 1), the OSC/12 signal is counted. The count is directly associated with the period that INT0 was high. In this situation, it would make some sense to initialize the counter to 0 before starting the measurement.

If we were trying to generate a predetermined time period, the counter would still be fed from the OSC/12 signal (Figure 6-7). The GATE configuration bit would be 0, and the TR0 configuration bit would be 1. We would like an interrupt to be generated after N counts, where N represents the number of counts associated with the time period. Because the counter is an up counter, we would like to preload the counter with the maximum count, minus these N counts.

As an extension of this time period concept, when the counter clock is periodic, it would be nice to enter the actual time period and have Excel generate the count required to achieve this.

For want of a better expression, I refer to these as "counting types," and these concepts can be extended to the other modes as well.

I have implemented these concepts using two Combo boxes (see Figure 6-13). The first of these two Combo boxes selects between the Counter and Timer concepts with these two options being listed in cells A10 and A11. The linked output cell is A9. The second of the two Combo boxes selects the Counting Type providing the options "Count from nn", "Count nn", and "Timer" to cover the concepts just described. The input cells are A13 to A15, and the output is in cell A12.

It is necessary to unprotect cells A9 and A12 because of the way Excel works with the output from controls. If protected and the whole sheet is protected, any event that attempts to change their value will generate an error message.

The configuration of this counter is achieved by writing to the four least significant bits of the TMOD register. The outputs of the first two Combo boxes are used to set the bits in the register representation, which I set up in cells B12 to F15, as seen in Figure 6-13. The gate function (cell B15) may be set or cleared by entering a 1 or 0 in the cell, so it must be unprotected. In addition, its value can only be 0 or 1 so the cell is formatted for zero decimal places, and also conditionally formatted. Click on the cell, and then **Format | Conditional Formatting** and choose the conditions of less than 0 or greater than 1 as in Figure 6-12. The background of the cell will turn to red if the incorrect value is entered.

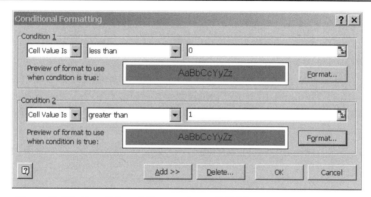

Figure 6-12: Conditional Formatting cell B15.

The other bits in TMOD are derived from the Combo box linked cells. The value of C/T (cell C15) is calculated from:

   =IF(CT0=1,0,1)

That is, if the value of linked cell "CT0" is 1, then this bit is a zero. If it is anything else (and the value is only set to go to 2), then the bit is 1.

Calculating M0 and M1 may not be as intuitive as you (certainly I) might think. I assumed that if I logically ANDed the value in the linked cell "TC0_Mode" with the number one, the result would be (as in programming) the least significant bit. This would look like:

   AND(TC0_Mode,1)

However the AND function assumes both the inputs are logical values, and if the value of TC0 is not zero, then it is TRUE.

I had to find another way to calculate bits M1 and M0. I opted for the CHOOSE function as follows. For M0 (in cell E15), I entered:

   =CHOOSE(TC0_Mode,0,1,0,1)

and for M1 (Cell D15), I entered:

   =CHOOSE(TC0_Mode,0,0,1,1)

Using concatenation and the dec2hex function, I created the hex equivalent of this nibble using the "n" character to remind the user that this is, in fact, a nibble and the most significant nibble is still to be added. Cell F15 contains:

   ="n"&DEC2HEX(B15*8+C15*4+D15*2+E15,1)&"H"

Figure 6-13 shows the result so far.

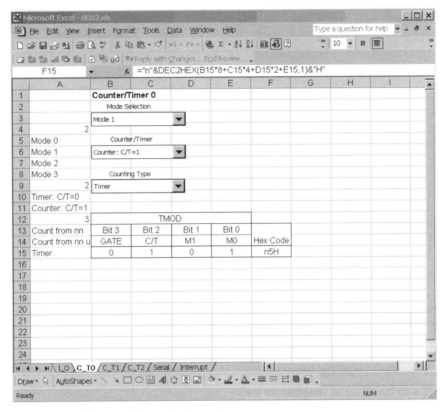

Figure 6-13: Setup of Counter Timer 0.

Timer Counter Control Register TCON

Only one bit in the TCON configuration register needs to be setup, and this is TR0, the Timer control bit. We will use a Check box to indicate if this is set or not. The linked cell is A17 and named *TR0*. The control is just visible at the bottom of Figure 6-16.

Counting Types

Since the requirements of each of the counting types are dissimilar, rather than create huge IF statements, we will create each condition separately, and then hide or unhide the associated series of lines. Unfortunately, because the controls hover above the lines, they cannot be made to disappear.

In all of the counting types we need to enter a number, but the number can be in hexadecimal or decimal, we place Option buttons that can be used for all configurations of the counting types. The buttons are linked to A20, which has been named *Base* and of course, is unprotected.

The "Count up from …" type is simple enough, as all the user has to do is enter a value, which will be written into the registers. There is a setup for each mode in this counting type in the cell block A23:H36 in Figure 6-16. In each mode, there are a different number of bits that can be written to the registers. In each of the input cells (remembering that from the user's perspective, only one of these setup registers will be visible), I have added a comment as to how many bits are permitted. In addition, the input cell also has conditional formatting that will turn the cell red for a number that is too large. The conditional formatting is set to formula as shown in Figure 6-14.

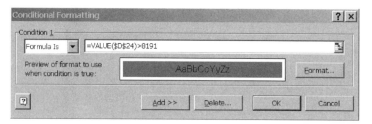

Figure 6-14: Using a formula in the Conditional Formatting box.

The data entry cell is formatted as text so that later manipulation is consistent and won't have to consider numbers as well as text.

To convert from text to a number, we use the VALUE function. Cell A24 will always have the decimal value of the number entered and it is translated to cell D25 so it can be entered into the source software for the project. Figure 6-15 is shown so that you are not totally in the dark. I have also used the **Tools | Formula Auditing** to indicate the precedents and descendents of the cells, in an attempt to clarify the sequence of events.

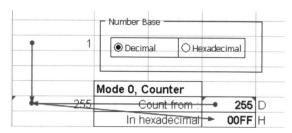

Figure 6-15: Initial register configuration on Mode 0, as a counter, and start counting from the designated value.

As before, the calculation is placed in column A to provide a focused area for the derived information and, of course, to hide it at a later stage.

In Parenthesis: *Excel Warning Detection*

Aside from the red triangles sitting in the top right-hand corner of a cell which indicates that the cell contains a comment, you may have noticed a green triangle in the upper left-hand corner of some cells. This is Excel's method of indicating that there may be a problem with the cell. Click on the cell with the green indicator, like A24, and you will notice an exclamation point in a diamond shape. Allow the cursor to hover over this symbol and Excel will provide a pop-up summary of what it believes to be noteworthy. Click on this symbol and there are a host of things that you may do associated with this perceived problem.

In each of the modes 1 to 3, while set as a counter for the counting type of "Count From", the register setting is an expanded form of the Mode 0 setup just described. TL0 and TH0 are treated as two separate registers. In Mode 3, TH0 does not work as a counter, so the user will have to manually do a timer calculation if TL1 is set as a counter. In all my years of experience, I have never used this mode so I have made an "authoritative" (pun intended) decision not to pursue this avenue.

Figure 6-16 shows the four modes so far. I have also added the precedents and descendents for TH0, since this is vaguely different to Figure 6-15.

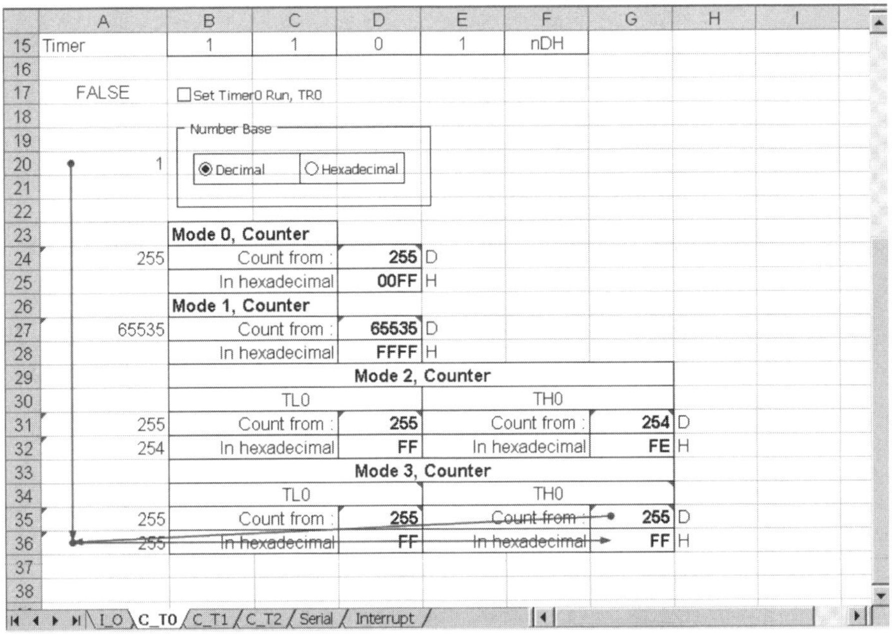

Figure 6-16: Modes 0–3 for Counter input, counter type "count from". In usage, only one of the modes will be visible at a time, depending on the mode.

Count

The next four modes are very similar to previous modes. The only difference is that for any value entered, the programmed value is subtracted from the (maximum count + 1) to allow for the overflow. The whole previous modes (A23 to G36) block is copied and the associated test in cells changed for "Count from:" to "Count:". The entries in column A are also modified. For instance, cell A38 contains:

=IF(Base=1,8192-VALUE(D38),8192-HEX2DEC(D38))

where the maximum count of a 13-bit counter is 8191.

Figure 6-17 shows the appearance of this small section.

	Mode 0, Counter					
7937	Count:	255	D			
	In hexadecimal	1F01	H			
	Mode 1, Counter					
65529	Count:	7	D			
	In hexadecimal	FFF9	H			
		Mode 2, Counter				
	TL0			TH0		
1	Count:	255		Count:	254	D
2	In hexadecimal	01		In hexadecimal	02	H
		Mode 3, Counter				
	TL0			TH0		
1	Count:	255		Count:	255	D
1	In hexadecimal	01		In hexadecimal	01	H

Figure 6-17: Modes 0–3 for Counter input, counter type "count".

Timer

Although the microcomputer peripheral may be configured as a counter, it can act as a timer for a periodic input on T0. The user needs to enter the frequency in hertz and the desired period in milliseconds. Excel then performs the calculation for the necessary divisor, and checks whether this is attainable.

Figure 6-18 shows the layout for Mode 0 Timer.

	Mode 0 Timer					
T0 Frequency (Hz)	12000	Overflow Period (mS)	18.3		Actual (mS)	18.33333
		Divisor	220	D	Error	0.18%
			1FED	H		
	13 bits maximum					

Figure 6-18: Setup of Mode 0 in a timer application.

The user will be expected to type the number in hertz for the input frequency, and the overflow period in milliseconds. The maximum input frequency on T0 is limited to the 8051 oscillator frequency/24 and this fact is commented in cell D52 (reading 12000) in Figure 6-18. The cell is also conditionally formatted, but there is a slight hiccup here. A conditional formatting statement cannot access a cell on a different worksheet, so I added the following formula in cell A2:

=I_O!F4*1000000

so that frequency does appear on this worksheet. The Conditional Formatting box is shown in Figure 6-19.

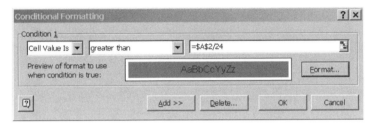

Figure 6-19: Conditional Formatting the input frequency.

The divisor is calculated from the input frequency divided by the output frequency and since the frequency is the inverse of the period, the contents of cell G52 are:

=D52*G52*1E-3

Since the divisor can only be set to an integer value, we need to introduce the ROUND function. So the cell becomes:

=round(D52*G52*1E-3,0)

In Parenthesis: *ROUND*

The action of the ROUND function is intuitive. Its format is *ROUND(number,num_digits)*.

Where *number* is the number to be rounded, and the *num_digits* is the number of decimal places. The function will round up at 1.50 and down at 1.49. A *num_digits* of zero will result in an integer, but of note is that you can use a negative number to move to the left of the decimal point. *ROUND (47.9,-1)* produces a value of 50.

This function should not be confused with the Number option in the cell formatting. In the ROUND function, the number is changed and any reference to the cell uses the rounded value. In the formatted state, the number is modified visibly, but the full accuracy is used and any further calculation.

In Mode 0 the counter is configured to count 13 bits, so the maximum it can be is 8191 and the cell is conditionally formatted to turn red if it exceeds this value. The formula in cell A53 subtracts the calculated divisor from 8192 (depending on decimal or hexadecimal format) since each counter is an up counter. The results appear as a hexadecimal number in cell G54. The cell at the bottom, reminds the user that the result can only be 13 bits long. This message changes to an alarm message if the result is greater than 13 bits.

Since the calculated divisor must be an integer, the resulting overflow is not always exact, so I added a calculation to the right, to substitute the divisor back and calculate the exact time and the percentage error.

The rows for Mode 1 are almost identical except that the counter is a 16-bit counter, so things are changed accordingly. Mode 2 is an 8-bit counter, so no further explanation is required.

Mode 3 results in 2 counters. The first is the same as Mode 2, but the second is clocked from the microcomputer oscillator (or at least 1/12 of it). It should be noted that since we cannot turn on a control, if the user desires to use the TH0 counter, they need to set the TF1 bit in the C_T1 worksheet.

To cut a long story short, the timer configurations are very similar to the counter. The configurations have been copied and modified to reflect that the input is from the oscillator/12. The "Count from nn" makes some sense if we are measuring a pulse width. Even though Mode 2 in this configuration is unlikely, it is still presented. "Count n" has the same meaning as the "Timer", and so the same configuration is used. Since we still have a lot of ground to cover, look at the workbook "8052.xls" if you are interested in further detail.

Macros to Hide and Unhide

As promised, we now consider hiding and unhiding lines, which we will use to show only one of the modes that we have developed. The easiest way to create the macro is to use the macro record facility, and then analyze and edit the result. It normally helps to think about the process we want to implement in advance so that it can all be included in one macro. We will want to invoke the macro from different sheets so we should start recording the macro from a different sheet to where we are going to hide the lines.

Click on the I_O sheet. **Select Tools | Macro | Record new macro...** and title it *HideLines* and then click on **OK.** All our actions will now be recorded. Click on sheet **T_C0.** Click on the row selection bar 23 and drag to line 25. Right-click on the selection and select Hide. Then **Select Tools | Macro | Stop Recording.** If the Macro toolbar is present, you can simply click on the Stop control.

Let's repeat the process for unhide and name the process *RevealLines.* Then we go to **Tools | Macro | Macros,** select either macro and **Edit** it. The result should be similar to the following:

```
Sub HideLines()
'
' HideLines Macro
    Sheets("C_T0").Select
    Rows("23:25").Select
    Selection.EntireRow.Hidden = True
End Sub
```

```
Sub RevealLines()
'
' RevealLines Macro
    Sheets("C_T0").Select
    Rows("22:26").Select
    Selection.EntireRow.Hidden = False
End Sub
```

It is obvious that the macros are procedures (also called *subroutines*) in Visual Basic. Procedures allow for passing values as parameters, so if we convert these we can generalize the procedures and call them from anywhere.

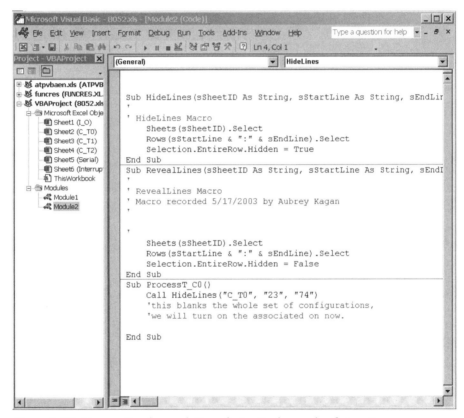

Figure 6-20: Preparation for revelation of a particular mode of operation.

Figure 6-20 shows how I have modified the procedures to allow for three strings (passed as parameters) that will allow for any set of lines on any sheet to be hidden. Once these parameters are added, the procedure (macro) cannot be accessed as a macro from any Excel feature like a control button or running a macro from the Macro list. These procedures have to be called from another macro, which you can see as ProcessT_C0. This is sufficient to test the process.

We can run the macro in the normal fashion, but in reality, we want the display to be updated every time we change the selection in a Combo box. Right-click on each of the Combo boxes and click on **Assign Macro**. Associate the control with the ProcessT_C0 macro that is listed in the window, as shown in Figure 6-21.

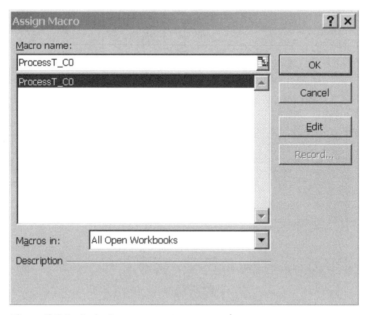

Figure 6-21: Assigning a macro to a control.

Once we have tested that the macro does indeed blank all the options, it is time to expand the scope.

Here is a snippet of the code that looks at all possible combinations and displays only one of the possible configurations. Each value of Mode Selection is treated as a case. Within each Mode case, there is a selection of a Counter or Timer case and that is further broken down into Counting Type cases. With the advantage of hindsight, it is possible to shorten the code by taking a different approach, but it may detract from the generality of the approach and would be more difficult to explain.

```
Sub ProcessT_C0()
    Call HideLines("C_T0", "23", "114")
    'this blanks the whole set of configurations,
    'we will turn on the associated on now.
    Select Case Range("A4").Value
    'choose based on "Mode Selection"
        Case 1
        'Mode 0
            Select Case Range("A9").Value
            'choose based on Counter/Timer
```

```
        Case 1
        'Timer
          Select Case Range("A12").Value
          'Choose base on Counting Type
            Case 1
            'count from nn
              Call RevealLines("C_T0", "76", "78")
            Case Else
            'timer & count n
              Call RevealLines("C_T0", "90", "94")
          End Select

        Case Else
        'counter
          Select Case Range("A12").Value
          'Choose base on Counting Type
            Case 1
            'count from nn
              Call RevealLines("C_T0", "23", "25")
            Case 2
            'count n
              Call RevealLines("C_T0", "37", "39")
            Case Else
              Call RevealLines("C_T0", "51", "55")
            'timer

          End Select

      End Select
    Case 2
    'mode 1
    ......
    ......
    Case 3
    'mode 2
    ......
    ......
    Case Else
    'mode 3
    ......
    ......
  End Select
  Range("B114").Select
  'deselect the range.
End Sub
```

Experimenting with the different choices in the Combo boxes hopefully will result in something similar to Figure 6-22. (I have hidden column A to improve the effect. I also removed the gridlines, formula bar and status bar.)

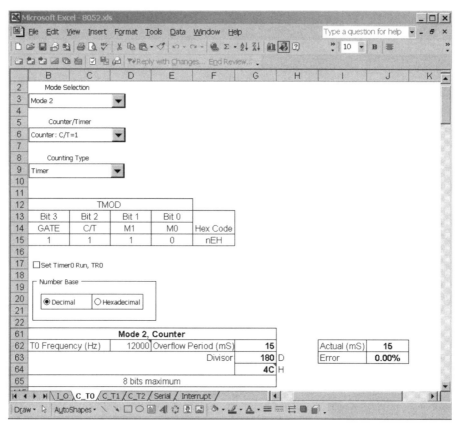

Figure 6-22: An operational worksheet!

Adding Forms

It seems to me that this application is missing a certain *je ne sais quoi*. Perhaps it lacks the verve and vitality of, say, a user manual. Seriously though, it would improve the interface significantly if there were graphic images of each configuration so that the parameters could be seen in context.

Based on the 8052 hardware description, I created the line drawings that I wanted using CorelDRAW® (of course any drawing software would work). I saved the output in several formats, but when I used the resultant graphics (coming up), the text in the graphics was unacceptable in every case but the Windows Metafile (.wmf) format.

In order to create a form we need to be in the Visual Basic Editor. **Insert | User Form** will result in Figure 6-23.

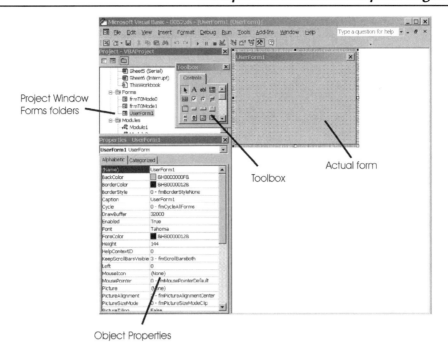

Project Window
Forms folders

Toolbox

Actual form

Object Properties

Figure 6-23: Adding a user form.

Note that the forms that are created appear as folders in the Project Window. When an object is selected, its properties can be viewed (if not immediately visible, right-click the object and select **Properties**). The properties can be viewed alphabetically or by category depending on the tab selected. We need to change some of these properties to suit our purposes. First, the name should be changed to provide a handle that is more meaningful. Click in the right-hand column of the (Name) row and give it a name. Visual Basic recommends that users implement "Hungarian" notation in naming objects. The recommended prefix for a form is "frm", so we name the form *frmT0Mode3*.

Change the **caption** (the text at the top of the window) to *Timer/Counter 0: Mode 3*.

In order to maintain the same size window for all the different graphics that will be displayed, change the **height** to 200 and the **width** to 260 (VBA will adjust these slightly at some point in the creation process). The form pops up over the worksheet and I didn't want it obscuring anything on the worksheet so that each time it changed, the user would have to drag it out of the way. I set up the window so that we manually place it at the top left-hand corner, so that the user could size the worksheet to fit in the right-hand side of the screen as shown in Figure 6-25. To this end, change **StartUpPosition** to **0-Manual** and **Left** and **Top** to 0.

If the modality of the window is true, the user must supply information or close the window to continue. In other words, with this property set to true, if the window pops up you are stuck there until it is closed. Since this is only a picture, we want the **ShowModal** property to be set to **False**.

As we will see shortly, the image that is displayed has a white background. For aesthetic reasons, I changed the **BackColor** (background color) to **white** (click on **Palette** tab and select the white color) so that the picture will appear to occupy the whole window.

Add Image Control

If the toolbox has gone missing, click on the form itself. If still not visible, click on **View | Toolbox.** Identify and click on the image control, and then click and drag a window on the form resulting in Figure 6-24.

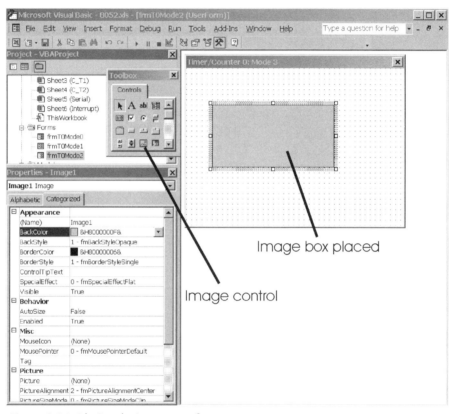

Figure 6-24: Placing the image on a form.

The properties of the image object must also be customized. First and foremost is obviously the image that is going to be displayed. Click on the **Picture** property and using the browse function, find the file that is to be used. In this case it is *T0Mode3.wmf*. Since we (at least I) want the picture to blend in with the whole window, the **BackColor** is set to **white**, and the **BorderStyle** is set to **0-fmBorderStyleNone**.

Scaling the image is mostly by trial and error. Set the **Autosize** property to **True** and the **PictureSizeMode** to **1-fmPictureSizeModeStretch**. Doing this results in the image borders falling outside of the form and so the edges are unreachable. Modifying the **height** and **width** properties to about 100 resizes the image so that it is manageable. The image is then sized by dragging the edges till it looks right.

Once all the forms have been created, the code must be altered to show and hide the forms. I first created a procedure to clear all the forms:

```
Sub HideForms()
    frmT0Mode0.Hide
    frmT0Mode1.Hide
    frmT0Mode2.Hide
    frmT0Mode3.Hide
End Sub
```

This is called at the beginning of ProcessT_C0, which you will recall is run every time one of the Combo boxes is modified. Depending on the setting of the Mode Combo box one of the forms is shown by using the associated instruction:

```
frmT0Mode0.Show
```

in each of the four possible cases. Running the spreadsheet and clicking on the mode box will result in a picture something like Figure 6-25.

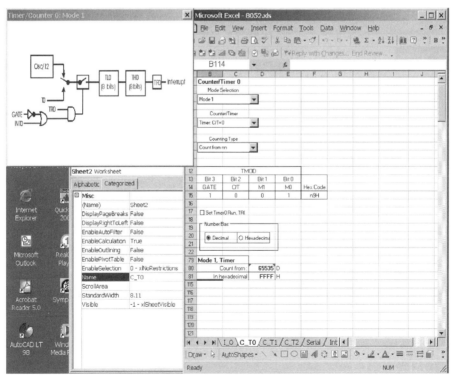

Figure 6-25: Running the application. The Excel window has been sized so that the form does not hide anything on the worksheet.

As with all programming, there is yet another issue. If we change the worksheet by clicking on a tab, the form does not disappear. We need to find an event (remembering that Windows is event driven) that occurs when the worksheet tab is clicked. In the project window, under the **Microsoft Excel Objects** folder is a **ThisWorkbook** folder. Double-click on this. In the left-hand drop-down box of the VB editor, select **Workbook**, and in the right select **Workbook_SheetActivate**. In this procedure, we add the code:

```
Call HideForms
```

This will resolve the problem as we can quickly test. However, we can even go one step further and use the change to detect which sheet is now valid and automatically initiate the form. The code becomes:

```
Private Sub Workbook_SheetActivate(ByVal Sh As Object)
    Call HideForms
    Select Case Sh.Name

      Case "C_T0"
          Call ProcessT_C0
    End Select
End Sub
```

And the process should act as expected. The only thing that does not happen, is when the workbook is first opened, no form is triggered. We will deal with that later after we have finished the whole application.

Timer/Counter 1 Sheet

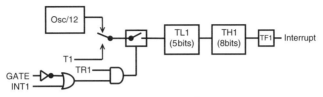

Figure 6-26: Timer/Counter 1, Mode 0.

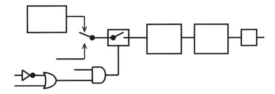

Figure 6-27: Timer/Counter 1, Mode 1.

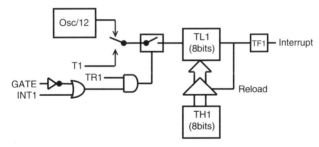

Figure 6-28: Timer/Counter 1, Mode 2.

The information on worksheet C_T1 is very similar to C_T0, as can be seen in Figure 6-26 to Figure 6-28. I simply copied the whole of page T_C0 to T_C1. The difference between the counters is that Timer 1 cannot have Mode 3, so all those options are deleted. When in Mode 2 it can be configured to operate as a baud rate generator.

The Mode Combo box is edited to have only three entries. As I have discussed earlier, it is not possible to modify the Forms controls, but it would be nice in this case since the baud rate generator only occurs in one combination of events. There is a way that is only slightly inelegant. The text to appear in the Counting Type Combo box appears on sheet C_T1 cells A13 to A16. A16, however, is conditional:

=IF(AND($A$9=1,$A$4=3),"Baud Rate","")

so that it can be blank or have text. The associated line on the drop-down box will either be a blank or carry the text "Baud Rate". It is still possible to click on the space, but the software in the macro looks after this.

Actually, the baud rate is affected by a bit in the PCON register, which I have added to the I_O sheet. The baud rate is calculated from $(2^{SMOD}/32)*$ Timer 1 Overflow. This can be re-written as $(2^{SMOD}/32)*((Osc/12)/N)$, where N is the divisor. Rearranging this to get N, given the Baud Rate:

$N=(2^{SMOD}/32)*(Osc/32)/BR$ and cell G86 implements this as

$=ROUND(((2^8052.xls'!SMOD)*D85)/(G85*32),0)$

The circumflex $\wedge$ is the exponent symbol. It is also possible to express this using the POWER function.

A new macro is created for revealing the associated lines, new images created, and the other procedures modify to take into account the changes when working with sheet C_T1.

Timer/Counter 2 Sheet

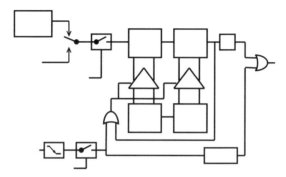

Figure 6-29: Timer/Counter 2, Auto Reload mode.

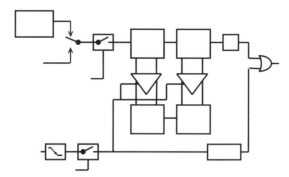

Figure 6-30: Timer/Counter 2, Capture mode.

The setting of the control bits for Timer/Counter 2 is completely different to timers 0 and 1 as can be seen in Figure 6-29 to Figure 6-31. There are three possible modes, but the Baud Rate mode can have three states: TCLK=1 and RCLK=0, RCLK=1 and TCLK=0, or TCLK=1 and RCLK=1. This is easily handled by just extending the mode selection.

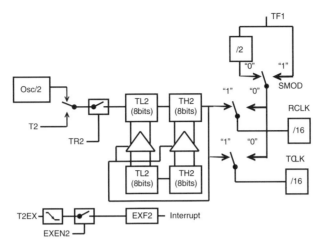

Figure 6-31: Timer/Counter 2, Baud Rate Generator mode.

The different calculations are very similar to Timer/Counter 0 and so I just copied the sheet and whittled it down. I also copied and modified the macros, along with creating new diagrams and forms. I won't detail much more of it, but if you want to see what all the formulas are, it is possible to see by going to **Tools | Options** and clicking on the **Formulas** box. You will see a screen similar to Figure 6-32 which will allow you to see which cells have formulas, and what they are.

In addition, you can also find which cells have been conditionally formatted. Follow the sequence **Edit | Go To | Special** and when the window as shown in Figure 6-33 is presented, select **Conditional formats** and then **OK**. This will result in all conditionally formatted cells being indicated.

Serial Port Sheet

Since the serial port is straightforward, the user interface needs no special mentions. No forms are associated with this worksheet.

Interrupt Control Sheet

The interrupt registers are straight forward and created in a manner similar to earlier worksheets as seen in Figure 6-34. The only item of note is that bits TR1 and TR0 in register TCON are derived from the settings on sheets C_T1 and C_T0 respectively. There is one user form associated with this worksheet. Obviously the macro "Workbook_SheetActivate" is modified to accommodate this.

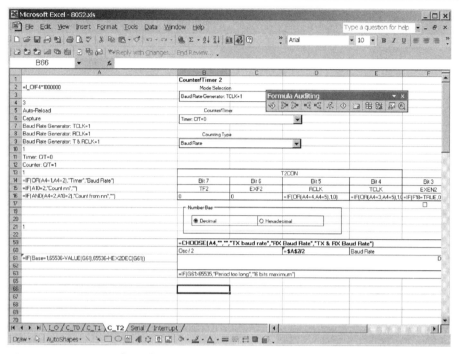

Figure 6-32: Viewing formulas.

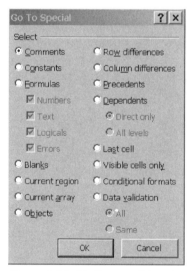

Figure 6-33: Finding conditional format in cells.

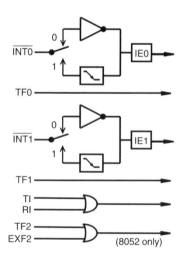

Figure 6-34: Interrupt configuration.

Summary Sheet

The summary sheet is where all the register data is accumulated in preparation to write it to a file. Some registers like SP or P1 are simply loaded from the hexadecimal representation in the associated worksheet. Registers like TCON are created by string manipulation to generate a 2-character hexadecimal number. Other registers like TH1 and TL1 are more problematic, because they depend on the mode of operation chosen. Typically, they are modified after the mode is fixed, so that the event that selects the mode cannot be used to write the values.

In order to solve this conundrum, we extract TH1 (or TH or TH2) and TL1 in each possible mode of operation. If the calculation results in a single 13 or 16 bit number then two bytes are extracted. The numbers are positioned in column L and M on each counter worksheet. They could be hidden at a later stage, but I have chosen not to do this.

Two cells in column A in each counter worksheet (remember they will be hidden later) are reserved. Each mode change uniquely triggers the revelation of several lines so the location of TH and TL can be isolated as well. The macro is therefore modified to write the cell identity where TH and TL are located into these two cells in column A.

For instance, on worksheet C_T0, these cell addresses are stored at A115 and A116. Each line revelation is now invoked by two calls. The first "unhides" the line, the second saves the cell locations to cells A115 and A116. The calling code has been modified to have this second call. Here is an extract from ProcessT_C0:

```
Select Case Range("A12").Value
'Choose base on Counting Type
   Case 1
   'count from nn
      Call RevealLines("C_T0", "76", "78")
      Call GenerateTHTL0("L78", "M78")
   Case Else
   'timer & count n
      Call RevealLines("C_T0", "90", "94")
      Call GenerateTHTL0("L93", "M93")
End Select
```

The routine that saves these values is short and sweet:

```
Sub GenerateTHTL0(sTH0 As String, sTL0 As String)
   'load the current pointers for TH0 and TL0
   Range("A115").Value = "C_T0!" & sTH0
   Range("A116").Value = "C_T0!" & sTL0
End Sub
```

The cells in the summary sheet use the INDIRECT function to get the cell address from A115 and A116 and then lookup the value. The sheet name must be added to the address in A115 and 116, otherwise the INDIRECT function will access the sheet that it is invoked on.

The idea of this model is to generate an assembler file with the necessary code to initiate the registers. Clicking on the button will run a macro that will do just that. The macro creates a text file from the name in cell B3. The assembly file consists of a series of MOV *regname,#data* type instructions. Depending on your assembler or even high level language you could create whatever file you wanted, following this example.

Initialize Values

By naming a macro *Auto_Open*, the macro is executed every time the workbook is opened. The Auto_Open macro in this workbook cycles through all the sheets setting up the default values and then positions the cursor at the location I think most users will start—the crystal frequency.

Conclusion

All through this exercise, we've gone to the trouble of unprotecting certain cells so that data can be entered and changed, but when we protect a sheet, we discover that the macros to hide and unhide will crash. The solution is to add an unprotect instruction for the sheet before a line is hidden or unhidden and to reinstate the protection after the action. For instance, the "HideLines" procedure becomes:

```
Sub HideLines(sSheetID As String, sStartLine As String, sEndLine As String)
' HideLines Macro
    Sheets(sSheetID).Select
    ActiveSheet.Unprotect
    Rows(sStartLine & ":" & sEndLine).Select
    Selection.EntireRow.Hidden = True
    ActiveSheet.protect DrawingObjects:=True, Contents:=True, Scenarios:=True
End Sub
```

All that is left to do is to hide column A in those sheets where it contains working data, protect the sheets and then go to **Tools | Options** and get rid of the **Formula bar**, **Status bar**, **Gridlines** and **Row & column headers**. The latter two have to be changed for each sheet. Figure 6-35 shows how well the application cleans up. You could almost forget it is Excel!

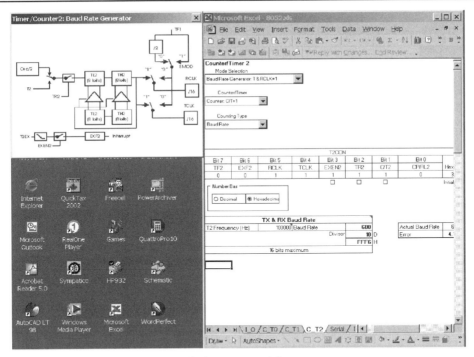

Figure 6-35: One of the sheets with the associated form.

EXAMPLE 7

Finding the Optimal Resistor Combination: LP 2951

Model Description

Many products in the electronics world rely on the ratio of two resistors to determine the output of a programmable device. This can include voltage regulators, amplifiers, and current sources. There is one degree of freedom, so the normal approach is to fix one of the values and solve for the second. This will produce a result, but because there are only discrete values for resistors, there is an inherent inaccuracy. We will see this in many forthcoming examples. However, sometimes we want to get as close as possible to a solution and that involves trying one resistor value against another for the optimal ratio.

The LP2951 is an adjustable voltage regulator. The output voltage is determined by the following formula:

$$V_{out} = V_{ref} \left(1 + (R1/R2) \right)$$

where R1 and R2 are the values of the feedback resistors as configured in Figure 7-1 and $V_{ref} = 1.23V$.

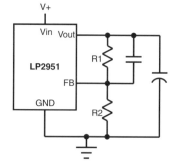

Figure 7-1: Programmable voltage regulator.

Custom Autofill

The flexibility in Excel allows the creation of custom lists for use with the Autofill feature. We electronic guys have our own special sequences like the standard resistor values. It would be nice to have these as a custom list. I have opted to use the "A" decade values for two reasons. First, if we really want accuracy, it would be better to use 0.1% tolerance resistors, and these are more readily available in these "A" values. Second, it is easier to show fewer values in the figures.

A custom list must be formatted as text since the autofill function could not tell the difference between a normal numerical sequence and the custom input if it were numerical. This is not a problem as we will see. In order to enter the series, we must first select all the cells that will be in the range and format them (**Format** | **Cells** | **Number** | **Text**) and then enter into those cells the sequence *10, 11, 12, 15, 16*, and so on to *100*. It is not possible to reverse the formatting order—that is, the numbers followed by formatting as text—because the custom list creation will not recognize this. As an alternative, you may prefer to enter an apostrophe ' before each number. Excel will place the little green triangle in the left-hand corner to suggest that perhaps we have erred in formatting a number as text.

In Parenthesis: *Communicating Custom Lists Between Different Computers*

Any custom list is created as part of Excel and is not carried with the worksheet. As a result, when you load the "LP2951.xls" workbook, the autofill list will not be present. To install the list, simply block cells E5 to AC5 and import the range as described in this example.

Click on **Tools** | **Options** | **Custom Lists** and **Import** the range we have just created as in Figure 7-2. Anytime we enter the sequence *'10,'11* or any other pair and then autofill, Excel will expand with the standard resistors values that we have setup. Note that the apostrophe *must* be used. Despite the fact that this is text, if employed in a formula evaluation, Excel is flexible enough to interpret it as a number.

In Parenthesis: *Data Tables*

The data tables approach in Excel can be applied in one or two dimensions. In a single dimension, the user prepares a list of values that can be used for the input variable in a formula, and the result of the formula calculated for each of these input values is produced in a parallel list. In a two-dimensional data table, it is possible to vary two input parameters to the formula. One variable is listed on the horizontal axis, while the second is on the vertical axis. The result creates a table using the horizontal and vertical values for each row/column intersection in the table. This approach is a variation of the scenario technique.

The single parameter data table can operate in columns or rows. In columns, the range of input values are in the first column, and the outputs are in the second. The formula must be placed in the cell above the first cell of the output column. Similarly for rows, the upper row is used for inputs, the lower for outputs and the lower cell to the left of the output row contains the formula for evaluation.

In a two-dimensional data table where the data is arranged horizontally and vertically, the formula is placed in the cell above the column data and to the left of the row data. This is the approach used in this model.

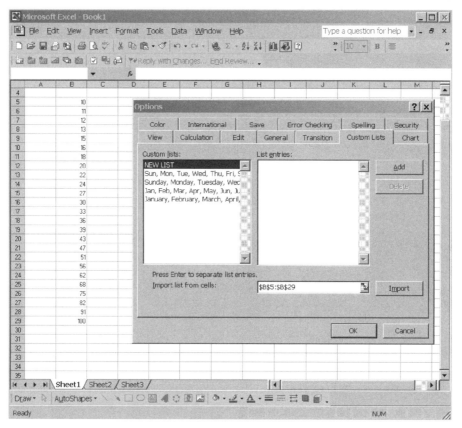

Figure 7-2: Creating a custom list.

Data Tables

Opening a new workbook, we create a cell at location C3 for the target voltage and name it *TargetVoltage*. Starting at cell E5, we enter '10 and '11 in F5. We then autofill this to the value 100 in cell AC5. These values will be used for R2 in the calculation. We could create a method of entering the standard values for R1 in the column that covers several decades, but in order to save time we can rewrite the output voltage formula to get an idea of the ratio between R1 and R2. This ratio can then be used to select the one or two decades of data needed. The TRANSPOSE function will then allow us to take the initial sequence for R2, multiply each entry in the input row by the factor and place the product in a corresponding cell in the output column.

At an output of 15V, R1 would be a factor of 10 greater than R2. I added a cell at D3 named *Factor* that will contain this ratio. Block from D6 to D30. Click in the formula bar and enter: =*factor*transpose(E5:AC5)*, followed by **<Ctrl>** + **<Shift>** + **<Enter>** to enter an array formula. (See Appendix A for more on Array Formulas.) The E5:AC5 region can be established in the formula through clicking and dragging when the opening bracket of the parenthesis is typed. Cells D6 to D30 now have values ten times greater than the horizontal row. In truth, getting the exact size of the target can be painful. It is easier to size it larger than will actually be needed, and then simply delete the cells with the error messages.

What we are trying to discover is what ratio provides a minimum error. The error is the target value minus the output voltage, i.e., TargetValue–(1.23(1+(R1/R2))). Setting this model up requires a dummy calculation. I selected cell B6 to represent R2 and B7 to represent R1 and I entered any value there.

Click on cell D5 and enter:

$$=TargetVoltage–(1.23*(1+(B7/B6)))$$

and the result is shown in Figure 7-4.

In Parenthesis: *Transposing Data*

In some cases (like this model), we would like to transform data arranged in columns to data in rows and vice versa. The technique we use is dependent on what we are trying to achieve. To merely copy the data, select the range in question and copy it to the clipboard (**Ctrl + C** or similar) in the normal fashion. Click on the first cell of the destination. Click on the menu sequence **Edit | Paste Special | Transpose** (see Figure 7-3) and voila, it is done.

Figure 7-3.

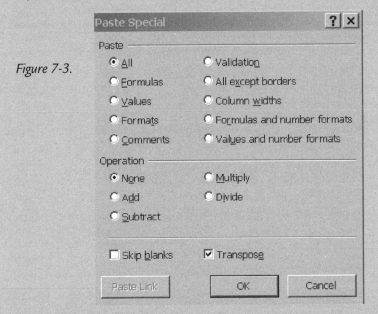

There is a formula that visually achieves the same effect, but it maintains the connection to the originating cell so that any value update in one of the original cells is reflected in the transposed cell. The relationship is not bidirectional. In order to transpose a range of data, first block the range of the destination, sizing it to the exact dimensions desired or larger. (Actually, if the exact size is not used, the function is still executed. If the area is too small, the data is truncated. If it is too large, error messages will occur in those cells.)

We enter:

=transpose (source range)

As we open the parenthesis, it is possible to click and drag the source range desired or to enter the top left and bottom right cell row/column identifiers. ***But, we do not press Enter***. Instead, we use a variation of the "three-fingered salute," **Ctrl + Shift + Enter**, and the transposition is done. This peculiar entry combination enters the formula as an "array formula." A convenient feature (used in this model) is that it is possible to perform a mathematical function on the original cells during the transposition.

This function is also available in VBA. In appearance, the execution of this function is similar to matrix manipulation and indeed it is. Those of you interested in other matrix functions should investigate MDETERM (matrix determinant), MINVERSE (inverse matrix) and MMULT (product of two arrays).

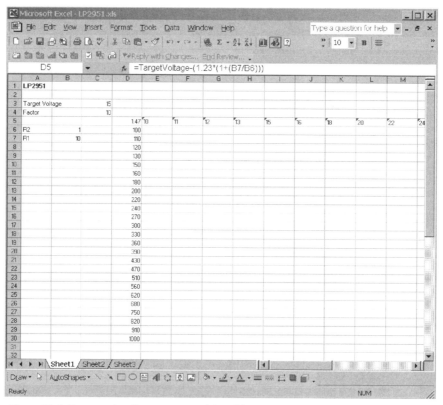

Figure 7-4: Preparation for a two variable table.

Block the range D5 to AC30, and then click on the menu sequence **Data | Table** and we are faced with the dialog box to associate the row and column data with the input data. Add the two cells used in the input calculation (see Figure 7-5) and click on OK. The area of the table is filled with data derived from the values in the respective row and column. Now would probably be a good time to manicure the table.

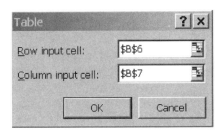

Figure 7-5: Associating row and column with input data.

Min Function

We are going to look at each column and try to find the minimum value in the column. In cell E31, we enter:

$=MIN(E6:E30)$

The minimum function does not return a value that is closest to zero, but the most negative number if it exists. In order to resolve this, we must edit the formula in cell D5 to:

$=abs(TargetVoltage-(1.23*(1+(B7/B6))))$

and we copy this to all the columns through to AC31. Row 31 now contains the minimum possible error in the output voltage for each possible value of R2. (The MIN function allows for up to 30 parameters, since it can act in an immediate sense, for example, $MIN(1,-4,5...)$, but a large-range input is considered a single parameter so it would be possible to enter the complete B series range. Or even on a particularly slow day, you could enter every resistor value from 1 ohm to 1M!

MATCH Function

In Parenthesis: *MATCH*

The function has the format:

MATCH(lookup_value,lookup_array,match_type)

This searches a lookup array for a particular value and returns the offset from the beginning of the range. The match type defines how the match is defined. For a match type value of 0, the function looks for an exact match. A match type value of –1 forces the function to identify the smallest value that is greater than or equal to lookup_value. Similarly, a match type value of 1 forces the function to identify the largest value that is less than or equal to lookup_value. In the case of –1 and 1, the array values must be ordered.

We also want to know (trust me for now) which row this minimum appears in. To find the row number, we need the MATCH function.

In cell E32 enter:

   =MATCH(E31,E6:E30,0)

and then copy this for all the cells E33 to AC33. Row 33 contains the row that will indicate which value or R1 gives the minimum error for each value of R2.

In order to find the minimum of the minima (that is the smallest error that can be found), enter the following in cell D31:

   =MIN(E31:AC31)

which searches all the minima for the smallest value.

In order to find the associated value of R2, we need to find the column in which the smallest error is found. In cell D32, enter the formula:

   =MATCH(D31,E31:AC31,0)

which identifies this column.

INDEX Function

In Parenthesis: *INDEX*

This function is formatted as follows:

INDEX(array,row_num,column_num)

It will return the value of the cell in the array at the intersection of the row and column chosen. It is possible to omit the row or the column for a single line range. An interesting option is to use the value 0 as a row, and the function will then return an array of all the cells in the column. The same is obviously true by interchanging the column and row. You will need to enter this as an array formula to make this option work.

In order to extract the resistor values for R1, we need to take the column number found in cell D32 and use it as the base of a lookup in the range identified as the range of resistor values (E32:AC32). We will extract the column number to using the INDEX function

Cell D33 contains:

   =INDEX(E32:AC32,,D32)

We could of course embed this as a nested function in the following lookups—it just seems easier to have it visible on the worksheet.

Finding the associate values is achieved using the index function again.

D34 has the value of R2 and contains the formula:

   =INDEX(E5:AC5,,D32)

while D35 has the value of R1 and contains the formula:

   =INDEX(D6:D30,D33,)

Obviously, the numbers are a ratio and can be scaled to the suitable value in K ohms as required.

Block Conditional Formatting

It is possible to emphasize the intersection where the minimum occurs by Conditional Formatting of the whole block. The do this, we click and drag across the whole range from D5 to AC30 and then click on **Format | Conditional Formatting**. Enter the condition for the rows and then **Add>>** the condition for the columns as in Figure 7-6.

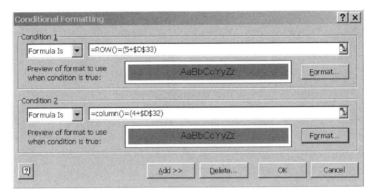

Figure 7-6: Block conditional formatting.

The ROW function returns the row number within the worksheet, so the formula takes into account the relative position within the range of the table, hence the (5+$D$33). The same is true for the COLUMN function. The bracket pair used in the conditions in Figure 7-6, (ROW( ) or COLUMN( )), is the notation used to indicate that the calculation pertains to each cell in turn and to use the associated cell/row number.

The completed application is seen in Figure 7-7. All the user has to do now is to enter the desired voltage and the factor needed in terms of the resistor ratio. Despite the brute force of the computer calculating every possible option, in my opinion this is quite an elegant solution to a problem that has faced most of us at one time or another.

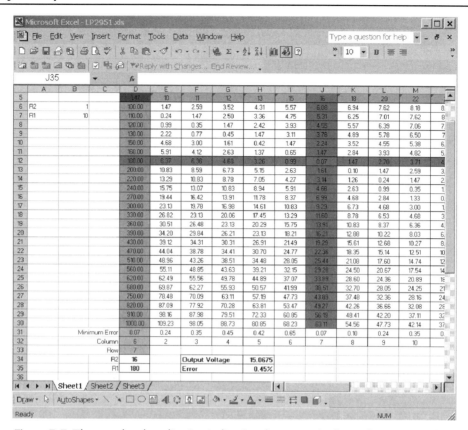

Figure 7-7: The completed application indicating the row and column that generated the smallest error and the resulting resistor ratio.

EXAMPLE 8

Resistor Color Code Decoder Using Speech Input

Model Description

A search on the web will reveal several approaches to entering the resistor color code on a computer and having the resistor value revealed. Doing this in Excel would not be that difficult using some Forms controls and a bit of code, but this would be reinventing the wheel. It seemed to me that whenever I was trying to decide what a resistor value was, both my hands were occupied. The simplest solution would be to say the colors out loud and have the computer decode this. Excel (or rather Office) 2002 has speech recognition built-in. Since modern resistors can have up to six bands and it is not always easy to tell which is the first or last band, an additional spin (pun intended) is to take the bands in as left to right, or right to left and look for a standard value that matches the pattern.

Resistors can have four, five or six color bands. In the early days when the color-coding was developed, the resistor value was represented by two color bands and one multiplication factor band. The fourth band represented the tolerance and would have one of three colors: nothing for 20% (hence the three-band resistor), silver for 10% and gold for 5%. As the resistor tolerance improved, a third band for the value was added and the tolerance colors were expanded to use the same color scheme as the value bands. More recently, a sixth band has been added for the temperature coefficient. These schemes can be seen in Figure 8-1, Figure 8-2 and Figure 8-3.

Figure 8-1: Three or four band marking.

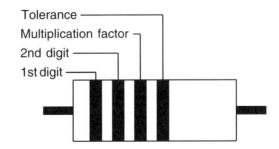

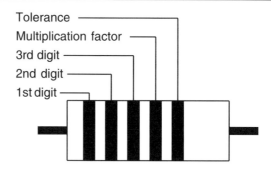

Figure 8-2: Five band marking.

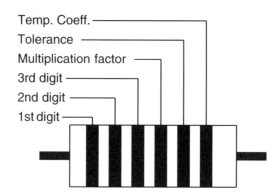

Figure 8-3: Six band marking.

The possible colors on a resistor and their associated values are detailed in the following table.

Color	Digits	Multiplication Factor	Tolerance	Temperature Coefficient
None			20%	
Silver	10*	0.01	10%	
Gold	11*	0.1	5%	
Black	0	1	1%	200 ppm
Brown	1	10	2%	100 ppm
Red	2	1K		50 ppm
Orange	3	10K		15 ppm
Yellow	4	100K		25 ppm
Green	5	1M	0.5%	
Blue	6	10M	0.25%	10 ppm
Violet	7		0.1%	5 ppm
Grey	8			1 ppm
White	9			
* needed in the application (see later)				

Let's see if we can generate a model that can handle all of this.

In Parenthesis: *Speech Recognition*

I used to regard speech recognition as a solution looking for a problem, but then I don't have a great reputation as a successful prognosticator. I remember seeing my first photocopy machine on an episode of "Mission Impossible" and wondering why anyone would ever want to copy anything. It was a much simpler time and place.

I am still not sure how practical this model will be in a noisy environment, but the feature is there and no doubt some of you will take the speech recognition into Excel applications I could never have imagined.

In Parenthesis: *Installing Speech Recognition*

Before we can use speech recognition, it must be installed. Open the Excel Help and search for "Speech Recognition". Locate the topic "Install and train speech recognition" and follow the instructions. The more you train the system the better the recognition gets.

Implementing Speech Recognition

The speech recognition function can operate in two modes: dictation and command mode. In the former mode, text is decoded from the speech and creates one long string. In the latter mode, any command from the toolbars can be invoked by speech. Selection between the two is through voice commands or clicking on the associated button in the language bar. If you have been through the voice training, you will have seen some of this in action.

It is possible to create voice commands by building a custom toolbar. On the toolbar we place buttons with the name of the word we will use as the command, like *Commence*. We will then create a macro with the same name as the command that will execute when the command is spoken.

I was faced with two possible approaches. I could have implemented a few commands and have the band colors dictated for later parsing. I opted for the second approach where each color acts as a command allowing for immediate feedback. There are twelve colors and three additional commands: "Commence", "Evaluate" and "Backup".

Quite a bit of preparation is required before the speech recognition can be implemented.

Viewing and Hiding the Language Bar

Working with the Language bar active can prove irritating, so while we set up the model, let's turn it off (if it is on). The easiest way to do this is to click on the microphone button of the expanded toolbar as seen in Figure 8-4(I), so that the toolbar collapses to Figure 8-4(II). (It is necessary to do this since the microphone can pick up and interpret signals even when the toolbar is minimized.) Then click on the minimize button on the Language toolbar and it is transformed into the EN (for English) icon in the system tray as shown in Figure 8-4(III).

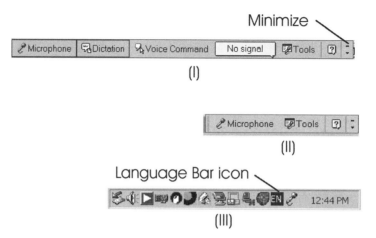

Figure 8-4: Language toolbar controls.

It can be restored by the menu sequence **Tools** | **Speech** | **Speech Recognition**, or by right-clicking on the Language Bar icon in the system tray and selecting **Show the Language bar**.

Worksheet Setup

First, we set up the worksheet as shown in Figure 8-5 to create a pictorial image of a resistor. The cell rows in the body of the resistor are merged in columns and sized so that it will be possible to generate a pictorial representation of the resistor. The body of the resistor is colored a light yellow so that a white band will show up against the light yellow background. Also, all the cells are center formatted.

Macros

We need to create four types of macros. The first, called *Commence*, will clear all the cells within the body of the resistor and place the cell selection on the first band. The second type is for all the different color possibilities. This macro will change the color of the cell selected, save the digit associated with the color within the cells allocated as the color band, and then move two columns to the right. A third macro, called *Evaluate*, will analyze the number sequence and pop-up the resistor value. Finally the fourth, titled *Backup*, allows the user to go back one cell and correct the color.

As I have mentioned before, the simplest way to start creating a macro is to execute the steps in the Record Macro mode and then edit them. For "Commence", I first ensured that the cell selected was outside the resistor body and then enabled the Macro Record. I clicked in cell C7 and dragged to P7, which selected the whole body of the resistor. I pressed the **Delete** key

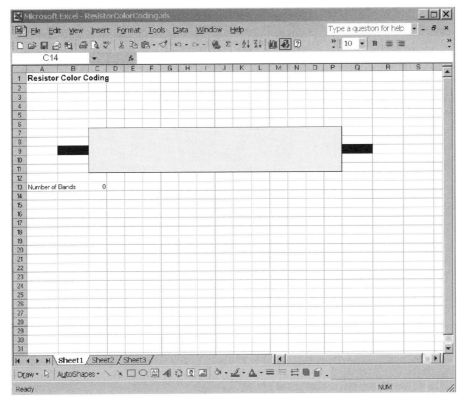

Figure 8-5: Resistor setup.

to remove any values, and then right-clicked on the selection and set the color to light yellow. This action cleared all of the bands. I then clicked on cell C13 and entered the value of zero, which zeroed the number of bands. Finally, I clicked back in cell D7 which positioned the cursor at the first band, ready for a new input.

This is the result and needs no further editing:

```
Sub Commence()
    Range("C7:P11").Select
    Selection.ClearContents
    With Selection.Interior
        .ColorIndex = 19
        .Pattern = xlSolid
        .PatternColorIndex = xlAutomatic
    End With
    Range("C13").Select
    ActiveCell.FormulaR1C1 = "0"
    Range("D7:D11").Select
End Sub
```

Let's create a typical macro to change the color of the band. Select a cell that contains a band and then start recording macro "brown". Enter the number "**1**", **Enter**, and press the arrow keys to get the next cell to the right. Terminate the macro to get the following result:

```
Sub Brown()
    Range("D7:D11").Select
    With Selection.Interior
        .ColorIndex = 9
        .Pattern = xlSolid
        .PatternColorIndex = xlAutomatic
    End With
    ActiveCell.FormulaR1C1 = "1"
    Range("F7:F11").Select
End Sub
```

There are some problems with this in making it generic. The initial range selection is not required so we should delete it. The second range selection at the end of the procedure is also limiting and we need to find an instruction that will move the active cell. Such a command would have the format:

```
ActiveCell.Offset (Rows,Columns).Range("A1").Select
```

We also need to keep a running count of how many bands have been entered. This tally is maintained in cell C13 (named as *NumberOfBands*), and is incremented by 1 every time a color macro is executed. In addition, there must also be a check so that there can only be a maximum of six bands. Since this increment algorithm is common to all the color bands, a subroutine would be in order. The "Brown" macro becomes:

```
Sub Brown()
' changes the cell color to brown and the value to 1
    With Selection.Interior
        .ColorIndex = 9
        .Pattern = xlSolid
        .PatternColorIndex = xlAutomatic
    End With
    ActiveCell.FormulaR1C1 = "1"
    Call MoveRight
End Sub

Sub MoveRight()
    If Range("NumberOfBands").Value < 5 Then
        'limit this to 6 bands
        ActiveCell.Offset(0, 2).Range("A1").Select
        'move selection 2 columns to the right
    End If
    If Range("NumberOfBands").Value < 6 Then
        'count up to 6
        Range("NumberOfBands").Value = Range("NumberOfBands").Value + 1
    End If
End Sub
```

We copy and paste and create macros for the remaining eleven colors. For the purpose of analyzing the colors later, silver has been assigned the value of 10 and gold the value of 11. The color value that is used is the index number in the color palette. How to get to these numbers is beyond me, so I recorded a macro that formatted a series of cells to the colors I wanted and I took those values and substituted them into the macros. These are the values that I used:

Color	Color Value	Numeric Value
Black	1	0
Brown	9	1
Red	3	2
Orange	45	3
Yellow	6	4
Green	4	5
Blue	5	6
Violet	7	7
Grey	15	8
White	2	9
Silver	16	10
Gold	44	11

For those of you who want greater adjustment, it is possible to give RGB values using the Color property.

The Backup macro moves left by two columns and formats the color back to the original light yellow. This is what it looks like:

```
Sub Backup()
    If Range("NumberOfBands").Value = 6 Then
    'last band and cursor hasn't moved to right so delete
    'this band first
        Selection.ClearContents
        With Selection.Interior
            .ColorIndex = 19
            .Pattern = xlSolid
            .PatternColorIndex = xlAutomatic
        End With
        Range("NumberOfBands").Value = Range("NumberOfBands").Value - 1
    End If

    If Range("NumberOfBands").Value > 0 Then
        'on the last cell the cursor does not move
        Range("NumberOfBands").Value = Range("NumberOfBands").Value - 1
        ActiveCell.Offset(0, -2).Range("A1").Select
        'move selection 2 columns to the left
```

```
        End If
        Selection.ClearContents
        With Selection.Interior
            .ColorIndex = 19
            .Pattern = xlSolid
            .PatternColorIndex = xlAutomatic
        End With
     End Sub
```

I am going to leave the Evaluate macro until later. You can run these macros at any time to see that they are functioning properly. Figure 8-6 indicates that they are!

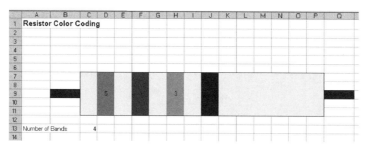

Figure 8-6: Checking that the macros work correctly.

To ensure that the workbook always starts in the correct location, I also added a macro called *Auto_Open*, which will run the "Commence" macro every time the workbook is opened.

Custom Toolbar

Excel allows extensive editing and creation of toolbars. To create a new toolbar we click on the sequence **Tools | Customize**, click on the **Toolbars** tab, and select **New**. You should be looking at something like Figure 8-7. Name the toolbar *Resistor*, and click **OK**. A small toolbar will appear on the screen.

Click on the **Commands** tab in the Customize dialog box. Scroll down and click on **Macros** in the Categories window. Click and drag the **Custom button** to the new toolbar and a smiley face appears as in Figure 8-8.

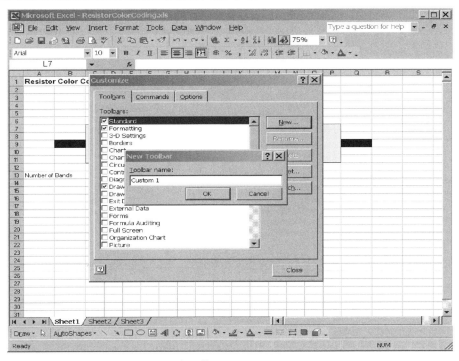

Figure 8-7: Creating a customized toolbar.

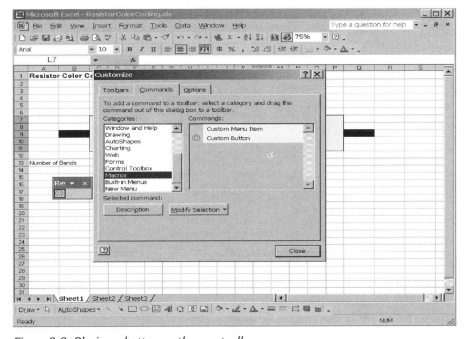

Figure 8-8: Placing a button on the new toolbar.

Right-click on this smiley and on the pop-up menu (Figure 8-9), change the **Name** to *Commence*, select **Text Only (Always)**, and **Assign Macro** assigning the button to the macro of the same name. Irritatingly, this process may take several right-clicks.

Figure 8-9: Changing button properties.

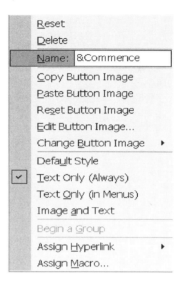

We repeat the process for all the colors and the Backup macros until we are left with a tool-bar that looks like Figure 8-10.

Figure 8-10: The Resistor toolbar.

Clicking on any one of the buttons should run the macro updating the bands on the resistor. We can get rid of the toolbar by clicking the "**X**" on the top right-hand corner, and make it reappear (or disappear) by right-clicking on a toolbar and selecting (or deselecting) the **Resistor** option. Of course, it is possible to delete entirely by going into **Tools | Customize | Resistor | Delete** but I don't think we should do that just yet.

> **In Parenthesis:** *Exporting a Toolbar*
>
> Toolbars are normally associated with an Excel installation. They can be inserted into a workbook for transportation, which is what I hope will happen with this example. In the customize dialog (Figure 8-7) having selected the **Toolbars** tab, click on the **Attach** button. In the next dialog box that appears, select the desired toolbar (in the "Custom Toolbars" panel) and click on the **Copy>>** button between the panels. The toolbar should be copied to the "Toolbars in Workbook" panel. Click on **OK**.

Adding Speech

We now re-enable the Language toolbar using the menu selections **Tools | Speech | Speech Recognition**. Ensure that the microphone is in the record state (also the mute button on the actual microphone, if there is one), click on the **Voice Command** button and then on the **Tools** button. On the drop-down menu, select **Add/Delete Words** and you will be presented with the dialog box of Figure 8-11. For each voice command that we are going to use, type the command word (like "Commence" or "Red") in the "Word" box, click on **Record pronunciation**, and annunciate the word. The software then adds the word to the dictionary. It is more reliable to record the exact word to associate with the macro than to rely on the speech recognition algorithm to recognize a word by application of rules.

Figure 8-11: Adding specific words to the word recognition database.

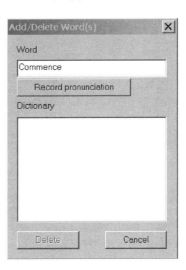

When all fifteen words (including Evaluate) have been added, close the box. It is time to try out how this works. Ensure that the Voice Command button is clicked on the Language toolbar and that the microphone is enabled. Now say the words and the macro should be executed. Simple enough!

By the way, as suggested by the Speech Recognition documentation, a good microphone really does improve the performance.

Evaluate the Color Code

The Evaluate function is far too lengthy to produce here in its entirety. It evaluates a different set of circumstances for each number of bands. An excerpt for the 4-band case is shown here:

```
Case 4
'4 bands
  nLegalValue = 0
  'preset value to indicate that the output is legal
  'if set to 1 then this is an illegal value
```

```
If Range("d7").Value < 10 Then
   nForward = Range("d7").Value * 10
   If Range("f7").Value < 10 Then
      nForward = nForward + Range("f7").Value
      If FindValueA(nForward) = 1 Then
      'found value
      Else
         nLegalValue = 1
      End If

   Else
   'second digit gold or silver
      nLegalValue = 1
   End If
Else
   nLegalValue = 1
   'to indicate an illegal value
End If
If nLegalValue = 1 Then
   Range("result").Value = "Forward value not found"
Else
   If Range("h7").Value = 10 Then
   'silver
      nForward = nForward * 0.01 & "R"
   Else
      If Range("h7").Value = 11 Then
      'golde
         nForward = nForward * 0.1 & "R"
      Else
      'any other value
         nForward = nForward * 10 ^ Range("h7").Value
         If nForward < 1000 Then
            nForward = nForward & "R"
         Else
            If nForward < 1000000 Then
               nForward = nForward / 1000 & "K"
            Else
               nForward = nForward / 1000000 & "M"
            End If
         End If

      End If
   End If
   'now for the tolerance
   Select Case Range("j7").Value
      Case 11:
```

```
                'gold
                    Range("result").Value = nForward & " 5%"
                Case 10:
                'silver
                    Range("result").Value = nForward & " 10%"
                Case 1:
                'brown
                    Range("result").Value = nForward & " 1%"
                Case 2:
                'red
                    Range("result").Value = nForward & " 2%"
                Case 5:
                'green
                    Range("result").Value = nForward & " 0.5%"
                Case 6:
                'blue
                    Range("result").Value = nForward & " 0.25%"
                Case 7:
                'violet
                    Range("result").Value = nForward & " 0.1%"
                Case Else:
                    Range("result").Value = nForward & " ??%"

        End Select
    End If

    'now for reverse
    nLegalValue = 0

    If Range("j7").Value < 10 Then
        nForward = Range("j7").Value * 10
        If Range("h7").Value < 10 Then
            nForward = nForward + Range("h7").Value
            If FindValueA(nForward) = 1 Then
            'found value
            Else
                nLegalValue = 1
            End If

        Else
        'second digit gold or silver
            nLegalValue = 1
        End If
    Else
        nLegalValue = 1
        'to indicate an illegal value
```

```
End If
If nLegalValue = 1 Then
    Range("result2").Value = "Reverse value not found"
Else
    If Range("f7").Value = 10 Then
    'silver
        nForward = nForward * 0.01 & "R"
    Else
        If Range("f7").Value = 11 Then
        'golde
            nForward = nForward * 0.1 & "R%"
        Else
        'any other value
            nForward = nForward * 10 ^ Range("f7").Value
            If nForward < 1000 Then
                nForward = nForward & "R"
            Else
                If nForward < 1000000 Then
                    nForward = nForward / 1000 & "K"
                Else
                    nForward = nForward / 1000000 & "M"
                End If
            End If

        End If
    End If
    'now for the tolerance
    Select Case Range("d7").Value
        Case 11:
        'gold
            Range("result2").Value = nForward & "  5%"
        Case 10:
        'silver
            Range("result2").Value = nForward & "  10%"
        Case 1:
        'brown
            Range("result2").Value = nForward & "  1%"
        Case 2:
        'red
            Range("result2").Value = nForward & "  2%"
        Case 5:
        'green
            Range("result2").Value = nForward & "  0.5%"
        Case 6:
        'blue
            Range("result2").Value = nForward & "  0.25%"
```

```
        Case 7:
        'violet
            Range("result2").Value = nForward & "  0.1%"
        Case Else:
            Range("result2").Value = nForward & "  ??%"
    End Select
End If
```

The procedure looks at the first two bands and if they are not gold or silver creates a number from the value stored within the band cell. The call to function "FindValueA" takes this number and compares it to all the legal numbers for resistors in the A series. If a match is found the function returns a value of 1, otherwise it is zero. If a match is found, the third band is used to scale the resistor value and present it in standard format (for example, 4.7 K), and then based on the fourth band the tolerance is tacked on to the result.

Since sometimes it is hard to figure out which is the first band and which is the last, the routine also reverses the process and evaluates the bands from right to left.

The results are saved in two cells reserved for the forward and the reverse readings.

Obviously, the interpretation of five- and six-band resistors has a few more lines of code and uses a different series of resistor values, but in principle, they work exactly in the same way.

Of some note in the software is the "exit for" statement. Those of you with "C" experience will know this as the "break" statement used with a "for" loop in order to break out of the loop. This is also the first time in this book we have used our own function call. A function only differs from a procedure in that it returns a value.

Now that the Evaluate macro is complete, we must add the button to the Resistor toolbar in exactly the same way as before using the **Tools** | **Customize** | **Commands** tab & **Macro** sequence, dragging the button to the toolbar and then changing the name and associating the macro. (There is a drop-down button on the Resistor toolbar that allows you to add a button instead of the above technique. It was possibly finger problems on my part, but I could not get this approach to work with speech recognition.)

Now we are almost ready to roll. You can try it out and see how it works. Obviously you don't need to use speech recognition; you can simply click on the toolbar buttons. The toolbar must be visible in order for the speech recognition to work.

Text to Speech

Not only do you get to talk to your computer, you can get your computer to talk back to you. You can change the properties of the speaking voice and the output device from the Speech icon in the Window Control Panel. Click on the **Text to speech** tab. Once this has been set to your satisfaction, return to the workbook.

Enable the Text To Speech toolbar by following the menus **Tools** | **Speech** | **Show Text To Speech toolbar**. Click on the **By Rows** button (as seen from the pop-up description) on the toolbar. Then block cells A15 to C16 and click on the **Speak Cells** button. The four cells should be read back to you.

It is simple enough to record this process to a macro called *Speak*, and the call to it is tucked in as the last thing to do in the Evaluate function. The only problem is that the Text to Speech function changes the active cell and that plays havoc with the backup function. We need to insert a method to record the current location and then restore it after the "Speak" procedure. We can do that using the following sequence:

> *vRow = ActiveCell.Row*
> *vColumn = ActiveCell.Column*
> *'saving current cursor location*
> *Call Speak*
> *'restoring cursor location*
> *Cells(vRow, vColumn).Select*

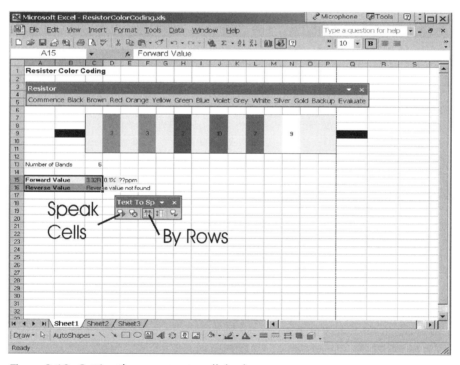

Figure 8-12: Getting the computer to talk back.

Conclusion

So there you have it. I hope this application is not anachronistic given that the industry is moving to surface-mount resistors. When I finally find a use for a particular tool, it becomes obsolete! Isn't that just the way of the world?

EXAMPLE 9

RTD to 4–20 mA Converter: XTR105

Model Description

Temperature is one of the real-world measurements that is required in electronics and especially in industrial control. There are many techniques to convert the temperature into an electronic format. One approach is to use a Resistance Temperature Detector (RTD), which consists of a wire with a resistance proportional to the temperature. Different metal alloys have different characteristics and each type is specified by the principal metal in the alloy and its temperature coefficients. The resistance R_T is approximated by the Callendar-Van Dusen equation:

$$R_T = R_0 + \alpha\, R_0\, [T - \delta(T/100 - 1) - \beta\, (T/100 - 1)(T^3/100)]$$

But normally, only the α coefficient is given and lookup tables are provided. From this equation it is obvious that the relationship between resistance and temperature is nonlinear.

One of the most common RTD types is made with platinum wire, with an α of 0.00385, which has a resistance of 100Ω at 0°C. RTDs are available in 2-, 3- and 4-wire types. The additional wires are used to null the effect of the resistance of the wires connecting the RTD to the electronics.

As discussed in an earlier example, the 4–20 mA current loop is very popular as a means of transmitting an analog signal around a factory floor because of its high noise immunity to electrically induced noise and its ability to power the sensor (hence the 4 mA offset) while measuring the signal. At the bottom end of the input range, the current through the loop driver is controlled to 4 mA, and it will increase to 20 mA at full scale input.

The RTD and current loop are so common that Texas Instruments/Burr-Brown manufacture an integrated circuit (XTR105) that does the conversion. A basic circuit can be seen in Figure 9-1.

The XTR105 provides two identical current sources to drive the RTD and a reference resistor R_Z. The difference in voltages developed by these currents is amplified and conditioned to generate the 16 mA range at the output. At the minimum input temperature R_Z should be equal to the RTD value so that the input voltage differential is zero. The upper value

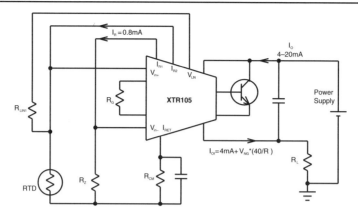

Figure 9-1: 2-wire RTD to 4-20 mA conversion.

is determined by the gain resistor R_G, and the XTR105 also has the ability to linearize the output with the addition of another resistor, R_{LIN1}.

The relationship between the resistor values is as follows:

$R_Z = R_{RTD}$ at T_{min}

$R_G = ((2R_1(R_2 + R_Z)) - (4(R_2R_Z)))/(R_2 - R_1)$

$R_{LIN1} = (R_{LIN}(R_2 - R_1))/(2(2R_1 - R_2 - R_Z))$

Where R_1 = RTD resistance at $(T_{min} + T_{max})/2$,

R_2 = RTD resistance at T_{max}

R_{LIN} = 1KΩ (internal to the XTR105).

Assuming we make this as a product where a customer can order any input temperature range, it would make an ideal model to implement in Excel.

Acquiring RTD Tables

The first step is to generate the RTD tables in Excel. After a search on the Internet, I accessed a table in HTML format from www.instrumentation.com (named for the company) for a platinum RTD, in degrees Celsius. In the browser, I selected **Edit | Select All** and copied the selection into a Wordpad file, where I gently massaged it and saved it as a text file which is on the CD-ROM as *table.txt*. An extract follows:

```
-200 18.52 -200
-190 22.83 22.40 21.97 21.54 21.11 20.68 20.25 19.82 19.38 18.95 18.52 -190
-180 27.10 26.67 26.24 25.82 25.39 24.97 24.54 24.11 23.68 23.25 22.83 -180
-170 31.34 30.91 30.49 30.07 29.64 29.22 28.80 28.37 27.95 27.52 27.10 -170
-160 35.54 35.12 34.70 34.28 33.86 33.44 33.02 32.60 32.18 31.76 31.34 -160
-150 39.72 39.31 38.89 38.47 38.05 37.64 37.22 36.80 36.38 35.96 35.54 -150
```

In order to get this into Excel, we follow the menu sequence **Data** | **Import External Data** | **Import Data**, and browse and select the "table.txt" file. We will be faced with Figure 9-2.

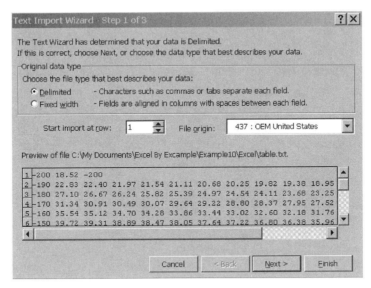

Figure 9-2: Importing a text file.

Ensure that the **Delimited** radio button is selected and click on **Next**, proceeding to Figure 9-3.

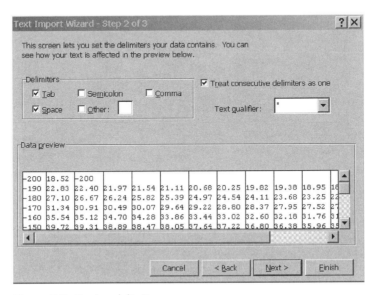

Figure 9-3: Setting delimiters.

Make sure the **Space** option is checked as a delimiter and click on **Finish**. The data will appear as in Figure 9-4.

	A	B	C	D	E	F	G	H	I	J	K	L	M
3	-200	18.52	-200										
4	-190	22.83	22.4	21.97	21.54	21.11	20.68	20.25	19.82	19.38	18.95	18.52	-190
5	-180	27.1	26.67	26.24	25.82	25.39	24.97	24.54	24.11	23.68	23.25	22.83	-180
6	-170	31.34	30.91	30.49	30.07	29.64	29.22	28.8	28.37	27.95	27.52	27.1	-170
7	-160	35.54	35.12	34.7	34.28	33.86	33.44	33.02	32.6	32.18	31.76	31.34	-160
8	-150	39.72	39.31	38.89	38.47	38.05	37.64	37.22	36.8	36.38	35.96	35.54	-150
9	-140	43.88	43.46	43.05	42.63	42.22	41.8	41.39	40.97	40.56	40.14	39.72	-140
10	-130	48	47.59	47.18	46.77	46.36	45.94	45.53	45.12	44.7	44.29	43.88	-130
11	-120	52.11	51.7	51.29	50.88	50.47	50.06	49.65	49.24	48.83	48.42	48	-120
12	-110	56.19	55.79	55.38	54.97	54.56	54.15	53.75	53.34	52.93	52.52	52.11	-110
13	-100	60.26	59.85	59.44	59.04	58.63	58.23	57.82	57.41	57.01	56.6	56.19	-100
14	-90	64.3	63.9	63.49	63.09	62.68	62.28	61.88	61.47	61.07	60.66	60.26	-90
15	-80	68.33	67.92	67.52	67.12	66.72	66.31	65.91	65.51	65.11	64.7	64.3	-80
16	-70	72.33	71.93	71.53	71.13	70.73	70.33	69.93	69.53	69.13	68.73	68.33	-70
17	-60	76.33	75.93	75.53	75.13	74.73	74.33	73.93	73.53	73.13	72.73	72.33	-60
18	-50	80.31	79.91	79.51	79.11	78.72	78.32	77.92	77.52	77.12	76.73	76.33	-50
19	-40	84.27	83.87	83.48	83.08	82.69	82.29	81.89	81.5	81.1	80.7	80.31	-40
20	-30	88.22	87.83	87.43	87.04	86.64	86.25	85.85	85.46	85.06	84.67	84.27	-30
21	-20	92.16	91.77	91.37	90.98	90.59	90.19	89.8	89.4	89.01	88.62	88.22	-20
22	-10	96.09	95.69	95.3	94.91	94.52	94.12	93.73	93.34	92.95	92.55	92.16	-10
23	0	100	99.61	99.22	98.83	98.44	98.04	97.65	97.26	96.87	96.48	96.09	0
24	0	100	100.39	100.78	101.17	101.56	101.95	102.34	102.73	103.12	103.51	103.9	0
25	10	103.9	104.29	104.68	105.07	105.46	105.85	106.24	106.63	107.02	107.4	107.79	10
26	20	107.79	108.18	108.57	108.96	109.35	109.73	110.12	110.51	110.9	111.29	111.67	20
27	30	111.67	112.06	112.45	112.83	113.22	113.61	114	114.38	114.77	115.15	115.54	30
28	40	115.54	115.93	116.31	116.7	117.08	117.47	117.86	118.24	118.63	119.01	119.4	40
29	50	119.4	119.78	120.17	120.55	120.94	121.32	121.71	122.09	122.47	122.86	123.24	50
30	60	123.24	123.63	124.01	124.39	124.78	125.16	125.54	125.93	126.31	126.69	127.08	60
31	70	127.08	127.46	127.84	128.22	128.61	128.99	129.37	129.75	130.13	130.52	130.9	70

Figure 9-4: RTD data loaded.

Note that for negative temperatures, the change in the values from left to right corresponds to the increase in the absolute value of the temperature and is inconsistent with the data presented for temperatures above zero in terms of a software lookup approach. I looked at tables provided by several RTD suppliers and quite a few seemed to use this approach. It is easy enough to use Excel to manipulate the data into a form that we need. Initially we need to mirror the data, so in cell N4 I entered the formula:

=L4

In O4, I entered the formula:

=K4

and so on, to:

=B4 in cell X4

I then blocked and copied N4 to X4 and pasted them into the range N5 to X23. Having done this, we no longer need the formula and we should revert to the values. Select the range N4 to X23 and copy it (**<Ctrl> + <C>** or through the menus). Then click on **Edit | Paste Special** and select **Values** (Figure 9-5).

Figure 9-5: Paste special, to convert formulas to values.

	K	L	M	N	O	P	Q	R	S	T	U	V	W	X
4	18.95	18.52	-190	18.52	18.95	19.38	19.82	20.25	20.68	21.11	21.54	21.97	22.4	22.83
5	23.25	22.83	-180	22.83	23.25	23.68	24.11	24.54	24.97	25.39	25.82	26.24	26.67	27.1
6	27.52	27.1	-170	27.1	27.52	27.95	28.37	28.8	29.22	29.64	30.07	30.49	30.91	31.34
7	31.76	31.34	-160	31.34	31.76	32.18	32.6	33.02	33.44	33.86	34.28	34.7	35.12	35.54
8	35.96	35.54	-150	35.54	35.96	36.38	36.8	37.22	37.64	38.05	38.47	38.89	39.31	39.72
9	40.14	39.72	-140	39.72	40.14	40.56	40.97	41.39	41.8	42.22	42.63	43.05	43.45	43.88
10	44.29	43.88	-130	43.88	44.29	44.7	45.12	45.53	45.94	46.36	46.77	47.18	47.59	48
11	48.42	48	-120	48	48.42	48.83	49.24	49.65	50.06	50.47	50.88	51.29	51.7	52.11
12	52.52	52.11	-110	52.11	52.52	52.93	53.34	53.75	54.15	54.56	54.97	55.38	55.79	56.19
13	56.6	56.19	-100	56.19	56.6	57.01	57.41	57.82	58.23	58.63	59.04	59.44	59.85	60.26
14	60.66	60.26	-90	60.26	60.66	61.07	61.47	61.88	62.28	62.68	63.09	63.49	63.9	64.3
15	64.7	64.3	-80	64.3	64.7	65.11	65.51	65.91	66.31	66.72	67.12	67.52	67.92	68.33
16	68.73	68.33	-70	68.33	68.73	69.13	69.53	69.93	70.33	70.73	71.13	71.53	71.93	72.33
17	72.73	72.33	-60	72.33	72.73	73.13	73.53	73.93	74.33	74.73	75.13	75.53	75.93	76.33
18	76.73	76.33	-50	76.33	76.73	77.12	77.52	77.92	78.32	78.72	79.11	79.51	79.91	80.31
19	80.7	80.31	-40	80.31	80.7	81.1	81.5	81.89	82.29	82.69	83.08	83.48	83.87	84.27
20	84.67	84.27	-30	84.27	84.67	85.06	85.46	85.85	86.25	86.64	87.04	87.43	87.83	88.22
21	88.62	88.22	-20	88.22	88.62	89.01	89.4	89.8	90.19	90.59	90.98	91.37	91.77	92.16
22	92.55	92.16	-10	92.16	92.55	92.95	93.34	93.73	94.12	94.52	94.91	95.3	95.69	96.09
23	96.48	96.09	0	96.09	96.48	96.87	97.26	97.65	98.04	98.44	98.83	99.22	99.61	100
24	103.51	103.9	0											
25	107.4	107.79	10											
26	111.29	111.67	20											
27	115.15	115.54	30											
28	119.01	119.4	40											
29	122.86	123.24	50											
30	126.69	127.08	60											
31	130.52	130.9	70											

Figure 9-6: Values converted.

Figure 9-6 shows the results to date. For the same selection, we cut (**<Ctrl>** + **<X>**) and paste it into the range B4 to L23 overwriting the original order. We block A3 to A22, cut it and shift it down a row. We delete the last two columns (L and M), add a little formatting and we are left with Figure 9-7.

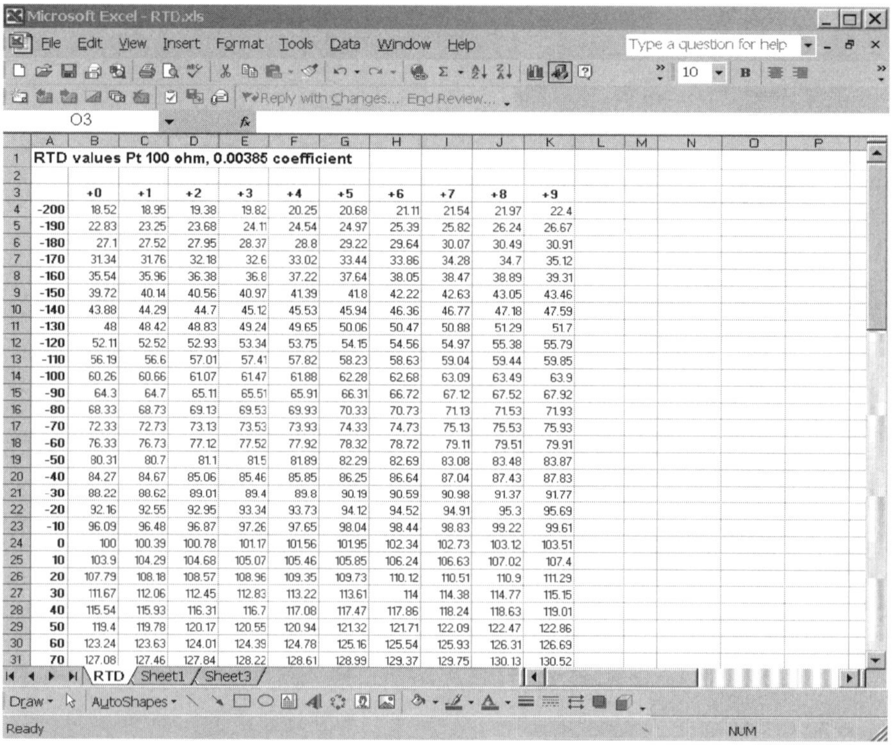

Figure 9-7: Completed RTD table.

Lookup RTD Value

The INDEX function has the format:

INDEX(array,row_num,column_num)

We have to manipulate the temperature to locate the correct row and column number. First, we enter any temperature in cell N1, just to start the process off. Let's use 125.

The rows increment by ten degrees, so we need to find the row based on the number of tens in the temperature. To do this we use the INT function. In cell N3, we enter:

=INT(N1/10)

and this returns a number of 12. The columns are based on the remainder of the above division. In cell O3 we enter:

=MOD(N1,10)

and it returns 5. We still need to do some manipulation of this. First, the table starts at –200°C, so we need to add a (200/10) for the offset and also the table (we will define later) starts at cell B4. For the INDEX function, row 1 column 1 defines cell B4, so we need to add a 1 to the 20 on the row offset and a 1 to the column offset to align the lookup action with the actual table.

If we put it all together, we enter in cell N5:

$$=INDEX(B4:K89,N3+21,O3+1)$$

We can play around with the value in cell N1 to see that the lookup works correctly.

Creating a Function

What we actually want to do is create a function with the temperature as an argument and the RTD resistance is returned as the value of the function.

As a programmer, you would immediately think of implementing the project using FOR loops to help identify the correct cell, but we have seen above that there is a perfectly good Excel function that can do the job. All we have to do is persuade VBA to use it.

Go to the VBA editor (**Macro** | **Visual Basic Editor** or **<Alt>** + **<F11>**). Insert a module (**Insert** | **Module**). Change the name of the Project to *RTDproject*, and the name of the module to *RTDmodule* in the module properties window. We need to do this because when this function is accessed from another workbook, the name must be unique or there may be a conflict. In the code window for the RTDmodule, add the code as shown below. The result appears in Figure 9-8.

```
Function RTDvalue(nTemperature As Integer) As Variant
    Dim nIntermediate As Variant
    Dim nIntermediate2 As Variant

    nIntermediate = Int(nTemperature / 10)
    nIntermediate2 = nTemperature - (10 * nIntermediate)
    nIntermediate = nIntermediate + 20 + 1
    nIntermediate2 = nIntermediate2 + 1

    RTDvalue = Application.WorksheetFunction.Index _
    (Range("[RTD.xls]RTD!b4:k89"), _
    nIntermediate, nIntermediate2)
    'add workbook reference to allow usage from
    'another module
    'Note the use of <space>_ to allow line
    'continuation
End Function
```

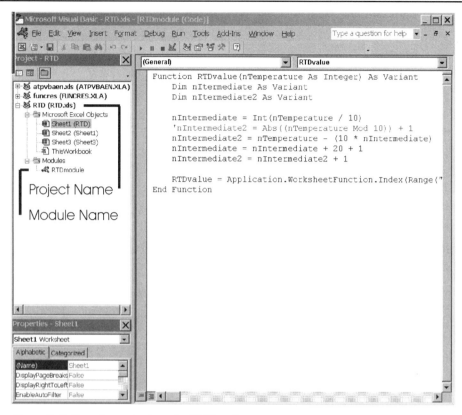

Figure 9-8: Function to lookup the RTD resistance.

The first part of the code recreates the Excel MOD function. The VBA Mod instruction does not perform in the same way and since they are named the same you cannot access the Excel function in this case. But VBA can access the Excel INDEX function using the Application. WorksheetFunction construct.

Accessing a Function

Although we will actually want to access the RTDvalue function from outside the module, this would be a good point to try and see that the function does indeed work.

On the Excel RTD worksheet, click in cell N7 and then using the menus, click on **Insert | Function** (Figure 9-9) and in the drop-down box select the **User Defined** category. Select the **RTDvalue** function and **OK,** and this leads to Figure 9-10.

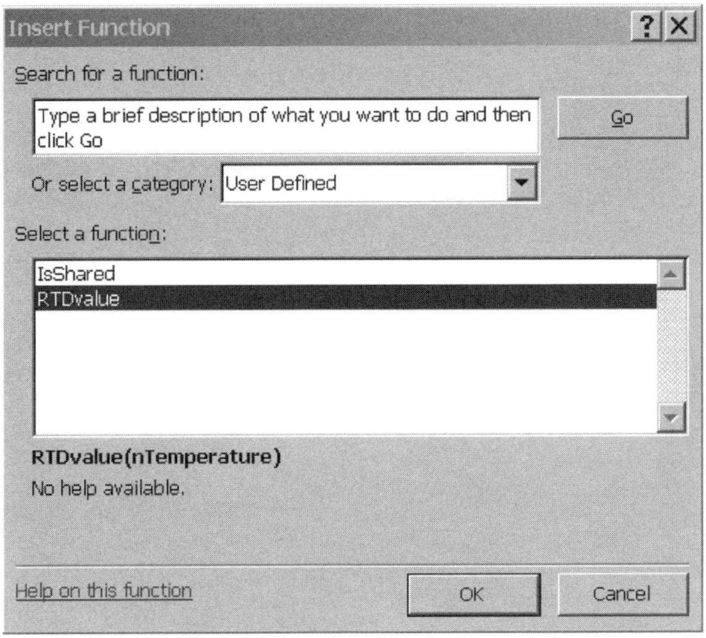

Figure 9-9: Insert a function.

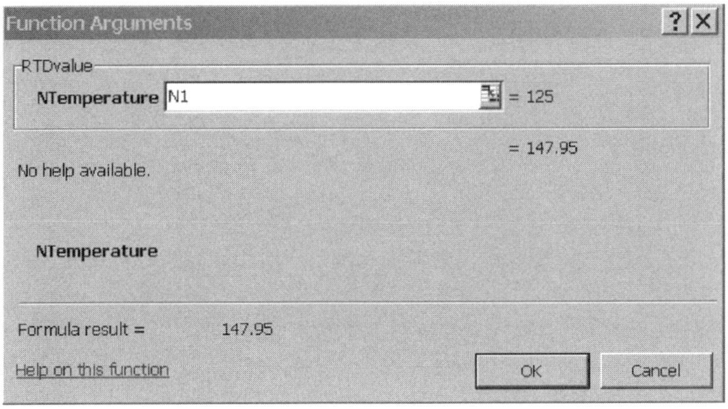

Figure 9-10: Selecting the input range.

Using the expand button, it is possible to actually click on the actual cell needed for the argument of the function, and click on **OK.** The cell now should have the same value as cell N5, which was calculated directly in Excel. Changing the temperature in N1 should lead to both cells N5 and N7 updating and showing the same value. Notice that there is a preview of the result in the lower part of Figure 9-10.

It is possible to avoid this Insert Function utility by simply entering the formula in F7:

$=RTDvalue(N1)$

Adding a Help Description to a Function

In Figure 9-9 and Figure 9-10, the comment "No help available" can be changed to provide something more informative. In the RTD workbook, go through the menus **Tools** | **Macros** | **Macros** (or <Alt> + <F8>). Only procedures will be found automatically, so we enter the name *RTDvalue* manually. We then click on the **Options** button and add descriptive text as in Figure 9-11. Click on **OK** and then **Cancel**.

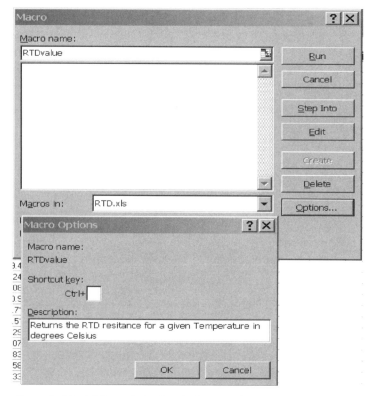

Figure 9-11: Adding a help description to a function.

Creating the Model in Excel

Open a new workbook and name it *XTR105.xls* using the **File** | **Save as** sequence. Open the VBA editor (<Alt> + <F11> or use the menus). In VBA, select **Tools** | **References** | **Browse** and select type of files to ***.xls**. Search for and select the "RTD.xls" files (Figure 9-12) and click on **Open**, followed by **Enter**.

We only need to do this if the worksheet containing the function (RTD.xls) is not open. Nevertheless, this will open the worksheet since we require the worksheet to be open in one form or another during regular operation.

Return to the new workbook and prepare the initial data, naming cells D4 to D7 for the descriptions in A4 to A7.

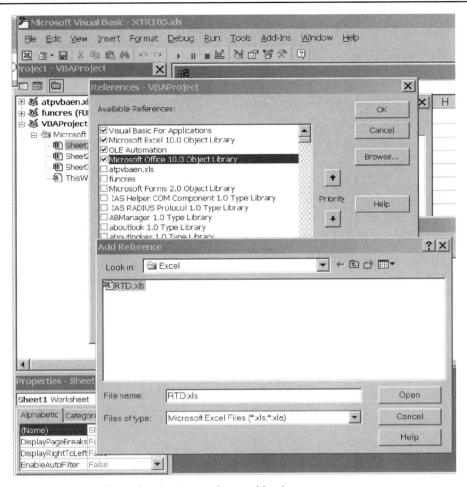

Figure 9-12: Finding a function in another workbook.

Figure 9-13 shows the formulas that are entered to calculate the resistor values (done by selecting the **Formulas** option in the **Tools | Options | View** sequence). The CONVERT function can be used to convert between many different kinds of units.

C27

	A	B	C	D
1	**XTR105 Resistor**			
2				
3				
4	**Reference Number**			7910000345
5	**Celsius/Farenheit**			F
6	**Minimum Temperature**			32
7	**Maximum Temperature**			500
8	Minimum Temperature (deg C)			=IF(CorF="F",CONVERT(MinTemp,"F","C"),MinTemp)
9	Maximum Temperature (deg C)			=IF(CorF="F",CONVERT(MaxTemp,"F","C"),MaxTemp)
10	R1	=RTD.xls!RTDvalue((degCmax+degCmin)/2)		
11	R2	=RTD.xls!RTDvalue(degCmax)		
12	RZ	=RTD.xls!RTDvalue(degCmin)		
13	RG	=(2*RE1*(RE2+REZ)-4*(RE2*REZ))/(RE2-RE1)		
14	RLIN1	=(1000*(RE2-RE1))/(2*(2*RE1-RE2-REZ))		
15				

Figure 9-13: Formulas behind the worksheet of Figure 9-14.

In Parenthesis: *CONVERT*

The convert function will allow a conversion from one measurement system to another. Its format is:

CONVERT(number,from_unit,to_unit)

There are many different conversions possible. To see them all, use the Excel Help feature. The units that will probably interest us are as follows:

Unit	Use as Argument
Weight	
Gram	"g"
Pound	"lbm"
Temperature	
Celsius	"C"
Fahrenheit	"F"
Kelvin	"K"
Distance	
Meter	"m"
Mile	"mi"
Inch	"in"
Foot	"ft"
Yard	"yd"
Angstrom	"ang"
Time	
Year	"yr"
Day	"day"
Hour	"hr"
Minute	"mn"
Second	"sec"
Power	
Horsepower	"HP"
Watt	"W"
Magnetism	
Tesla	"T"
Gauss	"ga"
Orders of Magnitude	
peta 10^{15}	"P"
tera 10^{12}	"T"
giga 10^{9}	"G"
mega 10^{6}	"M"
kilo 10^{3}	"k"
deci 10^{-1}	"d"
centi 10^{-2}	"c"
milli 10^{-3}	"m"
micro 10^{-6}	"u"
nano 10^{-9}	"n"
pico 10^{-12}	"p"
femto 10^{-15}	"f"

Figure 9-14 is the actual worksheet with the resistor results shown. The user is expected to choose Fahrenheit or Celsius, and the minimum and maximum temperature. The resultant values of R_Z, R_G and R_{LIN1} are produced in response to these inputs.

Figure 9-14: Initial worksheet.

	A	B	C	D
1	XTR105 Resistor Value Calculation			
2				
3				
4	Reference Number			7910000345
5	Celsius/Farenheit			F
6	Minimum Temperature			32
7	Maximum Temperature			500
8	Minimum Temperature (deg C)			0
9	Maximum Temperature (deg C)			260
10	R1	150		
11	R2	198		
12	RZ	100		
13	RG	212		
14	RLIN1	12277		
15				

Standard Resistor Values

Of course, you all know what the next step is going to be! Resistors are only made in discrete values and so we would like to know what values to use. Let's put the current project on hold while we investigate a new workbook (NearestValue.xls) that will allow us to look up the nearest resistor values. The functions return a numeric value rather than text so that they can be used directly in calculations. This is an interesting workbook since it does not need any entries on any of the sheets. It is purely an exercise in VBA programming. The four functions provided in the workbook allow for selection of resistors in the A (NearestValueA) and B (NearestValueB) series of values, for potentiometers (NearestPot), and a procedure that converts a number to the normal way of expressing the resistor value (LookupStandardResB).

By the way, if you don't want to get into the programming, the four functions can be accessed by simply using the module as an "add-in." You should then skip to the section titled, *Installing the NearestValues Add-In*.

There are several ways of generating a standard value for a given resistor value. The technique I adopted in the end, while it is a "brute force" approach, allows for simple expansion for other types of devices.

The code is too large to reproduce here, and anyway you have the source code on the CD, so I will show a snippet or two in order to elucidate.

The initial part of the function "NearestResistorA" defines the variables used. I found that I needed the double precision to prevent rounding errors in the value returned. The value under consideration, provided as an argument to the function, is called *CalculatedValue*. The first step in the process is to consider that the CalculatedValue must be greater than 1 Ω and less than 10 MΩ. If the value is outside this range, an error value is returned. In order for other Excel functions to interpret this as an error, the CVErr function must be invoked.

In order to simplify the identification of the lookup process, the value is changed to engineering notation. The mantissa is a variable titled *StandardForm* in the code, and the exponent is a variable called *Power*.

```
Function NearestResistorA(CalculatedValue As Double) As Double
    Dim i As Integer
    Dim Power As Long
    Dim StandardForm As Double
    Dim StdFrm As Double
    Dim Upper As Double
    Dim Lower As Double

    If CalculatedValue < 1 Or CalculatedValue > 10000000 Then
    'check for resitors < 1R or > 10M
        NearestResistorA = CVErr(xlErrValue)
    Else
        Power = 1
        StandardForm = CalculatedValue
        For i = 0 To 6
            If StandardForm >= 1 And StandardForm < 10 Then
                Exit For
            Else
                Power = Power * 10
                StandardForm = StandardForm / 10
            End If
        Next i
```

Using the mantissa as the key, we have a large Select Case statement. It caters for all possible ranges. On vectoring to any one of the cases, the standard value below the mantissa is stored on the "Lower" variable and similarly, the value above is stored on the "Upper" variable.

```
    Select Case StandardForm
        Case 1 To 1.1
        'set upper and lower to use later
            Lower = 1
            Upper = 1.1
        Case 1.1 To 1.2
            Lower = 1.1
            Upper = 1.2
        Case 1.2 To 1.3
            Lower = 1.2
            Upper = 1.3



        Case 9.1 To 10
            Lower = 9.1
            Upper = 10

    End Select
```

The next step is to consider which of the Lower or Upper values is closer to the mantissa and set the variable "StdFrm" to the complete value (that is the mantissa multiplied by the exponent):

```
If StandardForm - Lower > Upper - StandardForm Then
    StdFrm = Upper
Else
    StdFrm = Lower
End If
'use lower to save the value for later calculation
Lower = StdFrm
StdFrm = StdFrm * Power
```

In certain ranges, there are some values that are omitted so we need to consider this:

```
'now between 1 and 10 there are fewer values
If (StdFrm < 3.9) Or (StdFrm > 1000000) Then
    Select Case Lower
        Case 1 To 1.1
            Lower = 1
        Case 1.1 To 1.3
            Lower = 1.2



        Case Else
            Lower = Lower
    End Select
    StdFrm = Lower * Power
End If
```

And finally, the return value is set up and the function completed.

```
    NearestResistorA = StdFrm
    End If
End Function
```

The B range of values is implemented in exactly the same way except that of course there are significantly more values. The function is called *NearestResistorB*.

The lookup of potentiometers differs since the philosophy of using a variable resistor is different. You normally go for the value greater than the calculated value, but otherwise it is much the same. It is named *NearestPot*.

Finally, there is also a procedure that will take a number and format it to a standard resistor notation (as in 2.43 K ohms). It is titled *LookupStandardResB*.

Creation of Add-In

I will be using these functions in several models in this book, so it seems to me that it is a likely candidate as an add-in so that it will be readily accessible from any workbook. The inner machinations are irrelevant and using the add-in functionality, they can be hidden.

If it is not already open, open workbook "NearestValue.xls" and go to the VBA editor. Click on **Debug | Compile LookupValue**. Then right-click on the **LookupValue(NearestValue. xls)** entry in the VBA explorer bringing up Figure 9-15. Enter the information you want here. If you want, you can click on the **Protection** tab and fill in the required fields. Since you already have the source, this would be a pointless exercise for me.

It should be noted that by leaving this add-in unprotected, the code will be visible in all the workbooks (rather in the VBA environment) where this add-in has been enabled. This is not a problem, except that it is surprising when first noticed, and it is possible for you to change it (perhaps inadvertently).

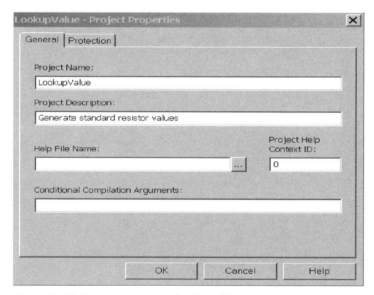

Figure 9-15: Setting properties for the add-in.

Now click on the sequence **File | Save As,** and save it as a **Microsoft Excel Add-In (*.xla)** type. This will automatically save it to the AddIns folder, but of course you can save it to anywhere you choose. I have added it to the CD-ROM as well.

Installing the NearestValues Add-In

Close the NearestValues workbook and open a new workbook. Click on **Tools | Add-Ins** and in the dialog box click on **Browse**. Locate the "NearestValue.xla" file and click on **OK**.

While the add-in has now been established for Excel as a whole, it is enabled or disabled in each individual workbook. As an example for the current model, open the "XTR105.xls" workbook as well as the "RTD.xls" workbook. Go through the sequence **Tools | Add-Ins**, and check the **NearestValue** option as in Figure 9-16.

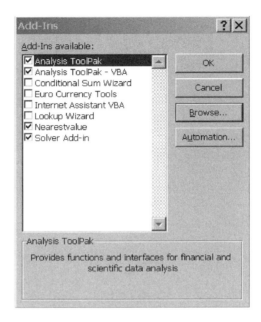

Figure 9-16: Installing an add-in.

Back to the Project At Hand

In the real world, component tolerances will lead to inaccuracies so each unit will have to be calibrated. To do this, a potentiometer should be placed in series with R_Z and a second one in series with R_G. This leads to four different values; two for the two fixed resistors and two for the potentiometers in series. Allowing for a 20% adjustment by the potentiometers (±10%), the result we calculate for R_{Zfixed} and R_{Gfixed} is 90% of the calculated value. We have set up the worksheet for the results to be entered in cells G4 to G8 (See Figure 9-19).

The easiest way to insert the function is almost identical to the procedure we used earlier. Click in cell G4 and click through **Insert | Function**. Select the **user defined** category, and then select the function NearestValueB as in Figure 9-17 and click on **OK**.

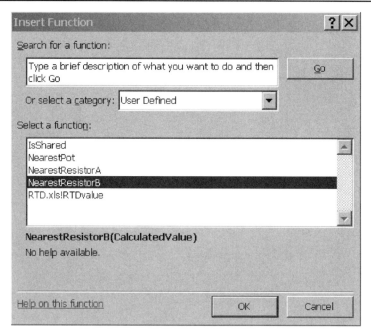

Figure 9-17: Inserting an add-in function.

The next dialog that pops up (Figure 9-18) is to provide the arguments for the function. The value entered is the calculation of 90% of the calculated value.

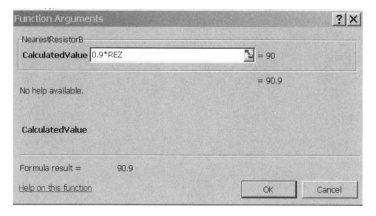

Figure 9-18: Adding an argument for the function. Note the real-time evaluation of the argument and the returned value of the function.

Click on **OK**. This procedure is followed for all four resistor values as shown using the formulas option shown in Figure 9-19, and the overall result is shown in Figure 9-20. There you have it—a fully functional workbook. But we can still make some improvements, provided you are prepared to join me in a little programming.

	E	F	G
1			
2			
3			
4		RZ fix	=NearestResistorB(0.9*REZ)
5		RZ pot	=nearestpot(0.2*REZ)
6		RG fix	=NearestResistorB(0.9*RG)
7		RG pot	=nearestpot(0.2*RG)
8		RLIN1	=NearestResistorB(RLIN1)

Figure 9-19: Actual values to be used.

	A	B	C	D	E	F	G
1	XTR105 Resistor Value Calculation						
2							
3							
4	Reference Number			7910000345		RZ fix	90.9
5	Celsius/Farenheit			F		RZ pot	20
6	Minimum Temperature			32		RG fix	191
7	Maximum Temperature			500		RG pot	50
8						RLIN1	12400
9	Minimum Temperature (deg C)			0			
10	Maximum Temperature (deg C)			260			
11	R1	150					
12	R2	198					
13	RZ	100					
14	RG	212					
15	RLIN1	12277					
16							

Figure 9-20: The worksheet to date.

Prompting for User Input

Let's assume that we produce this product with maximum customer flexibility, so they can call and order a product to measure the range from –17 to 213°C. Rather than tie up an engineer to calculate the values, we can create a macro for clerical staff to enter and produce the values required.

We need a user response to four questions: a reference number to tie the customer to the requested temperature range, a choice of Fahrenheit or Celsius, and minimum and maximum temperatures.

In Parenthesis: *InputBox*

VBA provides an easy way to prompt the user for input data. The format is

InputBox(prompt[, title] [, default] [, xpos] [, ypos] [, helpfile, context])

The only argument that is necessary is the prompt, which is the request posed to the user. It is possible to have several lines of text by concatenating the vbCrLf constant with the strings that make up the lines of text.

The title is the name that appears in the top left of the box. The default is the default string that you can use to suggest an input to the user. The box can appear anywhere on the screen using the xpos and ypos arguments. A helpfile can also be provided.

The InputBox function returns a string. This example shows how to check that the string is a number or a specific value.

The helpfile is the text file that contains the help for this function. It must have an input for the context, which vectors the help file to the correct entry.

There is also an InputBox method, which allows the box to return data types other than just strings. See the VBA help file for further information.

For the XTR105 workbook, I have added a module in the VBA Editor and then created a macro named *CreatePart*, which is a series of Input boxes and Message boxes to acquire the data.

It does little more than prompt the user for information and then checks that it is in the format needed. If the format is correct, the values are placed in the worksheet, and obviously a recalculation takes place. Take a look if you want to see how I used these interactive boxes.

Printout

If we want to print out part of a workbook, we first need to prepare for this by defining the print area beforehand. Click and drag the cursor over the areas of the workbook that are to be printed. They can consist of multiple areas created in the normal way of using the **<Ctrl>** key together with the mouse selection. Then follow the menu function **File** | **Print Area** | **Set Print Area** and Excel will then maintain a dotted line around the print area. I created it from cell A1 to G8.

As part of our ordering process, let's also assume that we would like to have a printout for the paper trail in production. Using the macro record function, we simply record the pressing of the print button (it uses the defined print area). Let's call this function *PartPrint*. It is also added as the last step of the CreatePart macro so that it prints out when the user has finished entering the data. Figure 9-24 shows the appearance of the project in operation.

In Parenthesis: *MessageBox*

The MessageBox function allows the program to announce something to the user. It also provides for a number of buttons so that the user can respond by clicking these buttons, and the function returns a number associated with that button. The format of the function call is:

MsgBox(prompt[, buttons] [, title] [, helpfile, context])

The only argument that is necessary is the prompt, which is the message displayed to the user. It is possible to have several lines of text by concatenating the vbCrLf constant with the strings that make up the lines of text.

The buttons to be shown are turned on by using VBA constants like vbYesNoCancel. It is possible to combine these with some form of special icon (which will carry an associated sound derived from the Windows setup), like vbQuestion. They are joined using the + symbol in a manner similar to this:

vbYesNoCancel+ vbQuestion

The title is the name that appears in the top left of the box.

The helpfile is the text file that contains the help for this function. It must have an input for the context, which vectors the help file to the correct entry.

Following is something that I found confusing. If you are not returning a value, that is you are treating the MessageBox as a procedure rather than a function, then you do not use the parenthesis in the code. For example, it would appear like this:

MsgBox "Procedure call", vbOKOnly as opposed to

Variable=MsgBox ("Function call", vbYesNo + vbQuestion)

Running Macros when the Workbook is Started

Irrespective of which of the two methods you used to reference "RTD.xls", every time you start "XTR105.xls" before opening "RTD.xls", you will be face with the pop-up window as shown in Figure 9-21. Even if we place the instruction to open the RTD workbook as part of the workbook startup event on "XTR105.xls", this message will normally show up.

On the assumption we would prefer a smooth user interface without this dialog, there are two possible approaches. First, you can set up "RTD.xls" to act as an add-in. This is quick and easy, but means that it is a module that is accessible to every workbook.

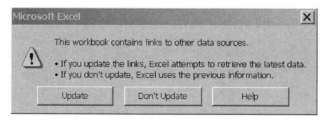

*Figure 9-21: Every time a workbook has an off book reference,
it starts by asking if the information should be updated.*

Second, I have not been able to find any documentation on it, so I had to go through a lot of trial and error, ending with infinite loops that needed the three-fingered salute (**<Ctrl>** + **<Alt>** + **<Del>**) to exit. I think I have the answer, but before you do try to run the macro, make sure everything is saved.

The actual change is very simple. Open the VBA explorer and click on the **ThisWork-book** folder of the VBAProject (XTR105.xls). In the **Properties** window, scroll down to **UpdateLinks** and select option 2 – **xlUpdateLinksNever**. Now save! This property is not available in Excel 97.

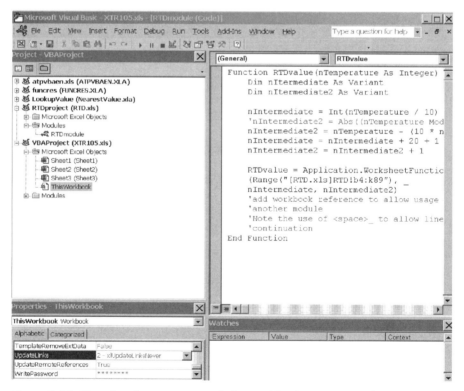

Figure 9-22: Changing the parameters of ThisWorkbook in XTR105.xls.

Double-click on the same "ThisWorkbook" folder and find the "Workbook_Open" event. This is the event that is triggered every time the workbook is opened. Add the text as shown in Figure 9-23. Be sure that the path in the Workbooks.Open line is correct.

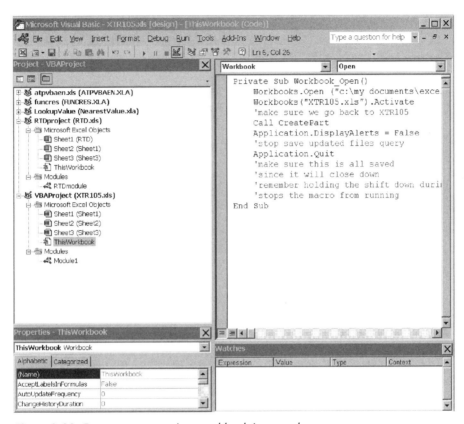

Figure 9-23: Run macro every time workbook is opened.

The first action that occurs is to load "RTD.xls". Because the CreatePart procedure is designed to operate with "RTD.xls" active, we must activate it first.

Also, we want the application to quit when we have generated the printout, but Excel will query if the workbook should be saved, so the statement *Application.DisplayAlerts=False* prevents that.

Every time the application is opened, the CreatePart macro is run, and at the end Excel is shut down—so remember to save the code before you run the macro! The macro also presumes that the user's printer is ready and waiting. The print request is handed to the operating system and spooling and errors are handled there.

In order to debug or change the workbook, open Excel and then while opening the workbook "XTR105.xls", hold the **<Shift>** key down. Also, remember the **<Ctrl>** + **<Break>** will also halt VBA program in its tracks.

Running from the Desktop

If we take this application to its logical conclusion, it is possible to create an icon on the desktop that, when activated (double-click), Excel is opened and the user is prompted for the inputs, the report is printed and Excel is terminated. Simply create a shortcut to "XTR105. xls" on the desktop. Change the icon and caption if you like. Now, the operator never has to be instructed how to start Excel, open a file and so forth.

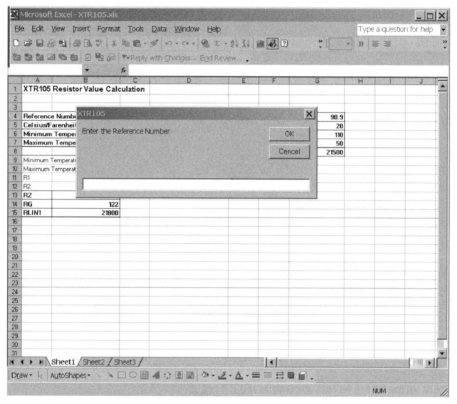

Figure 9-24: The application in full flight.

EXAMPLE 10

Voltage Regulator: LM317

Model Description

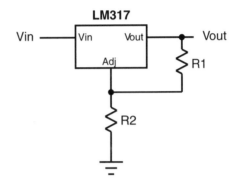

Figure 10-1: LM317 programmable voltage regulator.

The concept of using the ratio of two resistors to define a desired output pervades electronics. The LM317 was probably the first voltage regulator IC to implement it, but despite the age of the product, it is still in wide use. In any event, the techniques that I will use here are appropriate to all devices that rely on the resistor ratio approach. I will go on to consider heat sinking requirements as well, and this too may be extended to other applications

For the LM317 the output voltage is normally calculated by the first approximation equation, V_{out} = 1.25 (1 + (R2/R1)), since it is easy to rearrange the equation to find the resistor values for a given V_{out}. The equation is more correctly:

$$V_{out} = V_{ref} (1 + (R2/R1)) + I_{adj} R2$$

where V_{ref} is nominally 1.25V, and I_{adj} is the current flowing into the Adj terminal of the device.

Although in this example it may be possible to rewrite this formula in terms of R2, there are many equations where this is more difficult. As we all know, computers are very good at trying the same thing repetitively without complaint. Given a fixed value of R1, it is possible to try different values of R2 until the desired voltage is reached. Excel provides a function called *Goal Seek*, which does precisely that.

Installing the NearestValues Add-In

If you have not installed the NearestValues add-in, follow the instructions in the section titled *Installing the NearestValues Add-In* in Example 9. The functions in this add-in will allow us to look up standard resistor values.

Initial Model

We first create the model as shown in Figure 10-2. The values for V_{ref} and I_{adj} are derived from the data sheet for the LM317. We restrict ourselves to one degree of freedom and set the value of R1 to that suggested in the data sheet. Note the **formula** checkbox has been se-lected in the **Tools** | **Options** | **View** in order to show the formulas in the cells. Some cells have been suitably named.

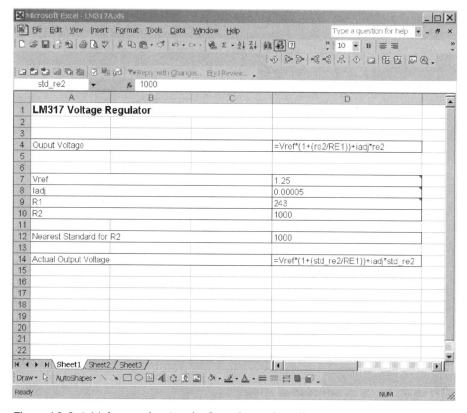

Figure 10-2: Initial setup showing the formulas in the cells.

The value for R2 and the nearest value for R2 are arbitrary at the moment, simply to setup the model. The output voltage is recalculated using the standard value as the last step.

Click in cell D12, and then follow the sequence **Insert** | **Function** and then select the op-tions as shown in Figure 10-3.

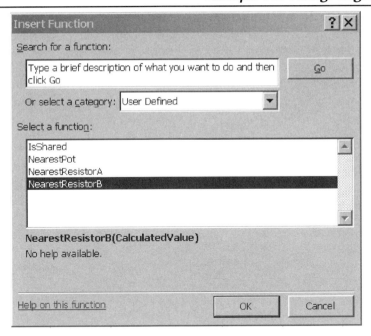

Figure 10-3: Inserting a custom function (in this case, the module is an add-in).

Click on **OK** and select the input parameter as in Figure 10-4.

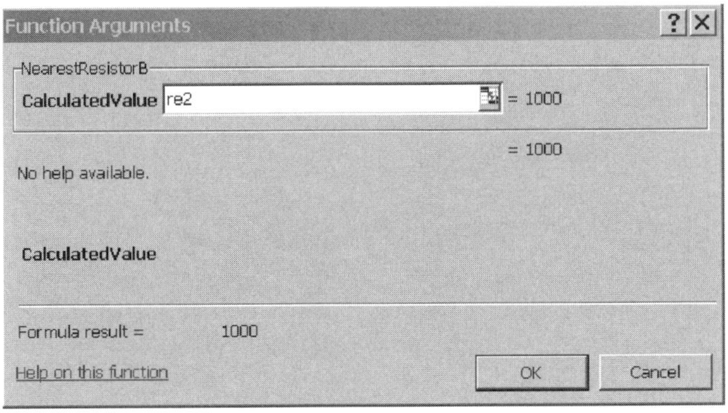

Figure 10-4: Selecting the input parameter for the function.

Alternatively, if you have a good memory you can enter:

=NearestResistorB(re2)

without the help features of Excel.

Goal Seek

Click on **Tools | Goal Seek** and enter the relative cells as in Figure 10-5, and then click on **OK**. Excel will calculate the value that solves the requirement, and the NearestResistorB function finds the nearest standard value. The Goal Seek function provides a summary as in Figure 10-6.

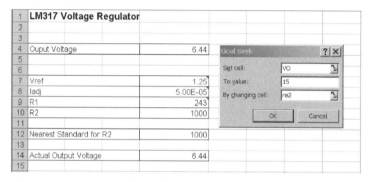

Figure 10-5: Priming the Goal Seek function.

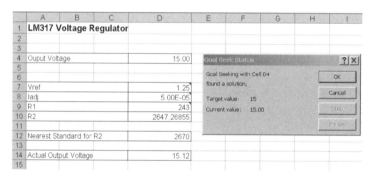

Figure 10-6: Summary of Goal Seek function.

Rather than go through the Goal Seek sequence every time, a macro could save the drudgery of repeating the actions.

As we have done before, we record the macro and then modify it. Follow **Tools | Macro | Record new macro** and set the macro as in Figure 10-7 (naming the macro *LM317*), and then follow the same sequence to Goal Seek a value (any value) and then stop the recording.

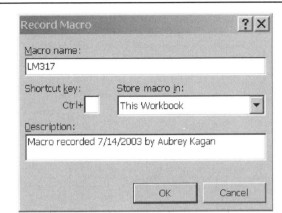

Figure 10-7: Recording a macro.

The recorded macro is:

```
Sub LM317()
    Range("D4").GoalSeek Goal:=10, ChangingCell:=Range("D10")
End Sub
```

We now need to add some functionality to prompt the user for the desired output voltage. To do this we need the InputBox function, which we add as the first step of the macro as follows:

```
Sub LM317()
    Dim nNewGoal As Single
    nNewGoal = Val(InputBox("Enter the output voltage", "LM317 Output Voltage Configuration", "5"))

    Range("D4").GoalSeek Goal:=nNewGoal, ChangingCell:=Range("D10")
End Sub
```

Figure 10-8 shows how the different arguments for the input box are presented.

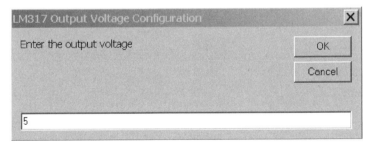

Figure 10-8: Input box showing the prompt, title and default values.

Worst Case Analysis

It is a simple matter to add a worst case analysis to find the possible variation range of the output voltage based on the tolerances of the resistors and the LM317. This is at 25°C. Figure 10-9 shows the formulas that are used, and Figure 10-10 has been reformatted to show the actual appearance.

	Nominal	Min	Max
3			
4	=Vref*(1+(re2/RE1))+iadj*re2		
5			
6			
7	1.25	1.2	1.3
8	0.00005	0	0.0001
9	243	=0.99*RE1	=1.01*RE1
10	3032.32579328923		
11			
12	=NearestResistorB(re2)	=0.99*std_re2	=1.01*std_re2
13			
14	=Vref*(1+(std_re2/RE1))+iadj*std_re2	=Vrefmin*(1+(RE2min/RE1max))+Iadjmin*RE2min	=Vrefmax*(1+(RE2max/RE1min))+Iadjmax*RE2max

Figure 10-9: Formulas used in worst-case analysis.

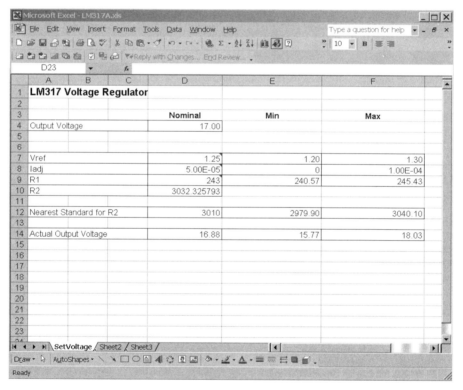

Figure 10-10: Appearance of the first sheet.

Let's add a Command button to the sheet to run the LM317 macro. Open the forms toolbox by clicking through **Insert | Toolbars | Forms**. Click on the Command button on the toolbox, and then click and drag a button on the worksheet and assign a macro when the dialog shown in Figure 10-11 appears.

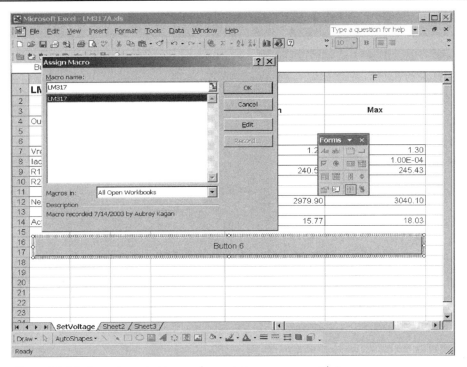

Figure 10-11: Inserting a button and associating a macro with it.

Click on **OK,** and right-click on the button to edit the button text to "Set Voltage". Click away from the button to end the change. Clicking on the button will now run the macro.

Thermal Analysis

Let's add another dimension by considering the thermal implications of the design. First, let's consider some theory.

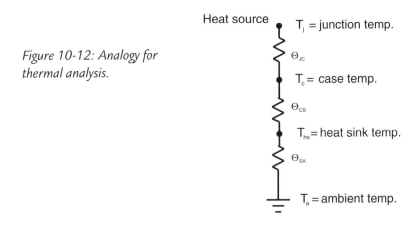

Figure 10-12: Analogy for thermal analysis.

It is possible to draw an analogy between thermal and electrical conductivity. The temperature corresponds to voltage, the thermal resistance to electrical resistance and heat flow to current. Using this approach and applying it to Figure 10-12, we can write

$$\Theta_{JA} = (T_j - T_a)/P_d \tag{1}$$

where Θ_{JA} is the thermal resistance from semiconductor junction to the ambient temperature (in °C/W), T_j is the junction temperature, T_a is the ambient temperature and P_d is the power dissipated.

$$\Theta_{JA} = \Theta_{JC} + \Theta_{CS} + \Theta_{SA} \tag{2}$$

where Θ_{JC} is the thermal resistance from the junction to the case, Θ_{CS} is the thermal resistance from the case to the heat sink, and Θ_{SA} is the thermal resistance from the heat sink to the ambient air.

The power dissipation, P_d, is calculated from the volt drop across the device V_d and the current flowing into it I_{in}.

$$P_d = V_d * I_{in} = (V_{in} - V_{out}) * I_{in} \approx (V_{in} - V_{out}) * I_{out} \tag{3}$$

The last approximation is true only where the quiescent current of the device is small in comparison to the output current.

In every case where heat dissipation is an issue, we must first consider the total power dissipation and provide enough heat sinking that is necessary to limit the junction temperature to a safe maximum. We need to consider the worst case of an application, and that may include a dead short across the output of the device.

Using these generalities with our specific example of an LM317T (that is the TO-220 package), the absolute maximum for the junction temperature is 150°C and traditionally we limit it to 25°C less than this. The next step in this example is creating a model that produces the required thermal resistance of the heat sink required.

Moving to Sheet2 and renaming it *Thermal*, we create the initial format as in Figure 10-13.

Each variable input can have several possible sources of data or value. For instance, as we shall see, the source voltage could be from DC or rectified AC. We are going to handle these alternatives by means of Option buttons.

The Option buttons are grouped together to deal with a single common aspect of the model. Each group of Option buttons is associated with a cell in column A, a column that we will hide later. The value of the cell corresponds to the Option button selected.

Depending on the design, the input voltage to the regulator can come from a DC source or some form of AC waveform. We will deal with a non-DC input later, but for the moment, the input voltage to the regulator will be derived from cell D3 for a DC input, or D4 for a rectified AC input, depending on an Option button selection. We first need to get the cor-

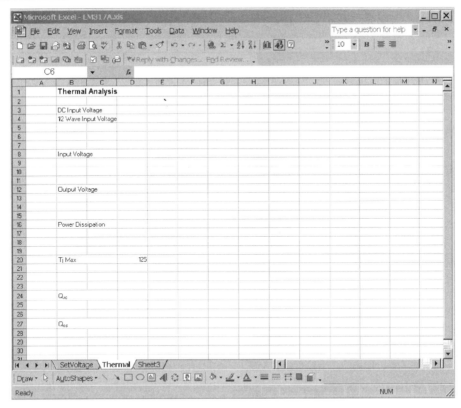

Figure 10-13: Preliminary setup for thermal analysis.

rect toolbox by clicking on **View** | **Toolbars** | **Forms**. Then click on the **Group box** icon and then click and drag an area as in Figure 10-14.

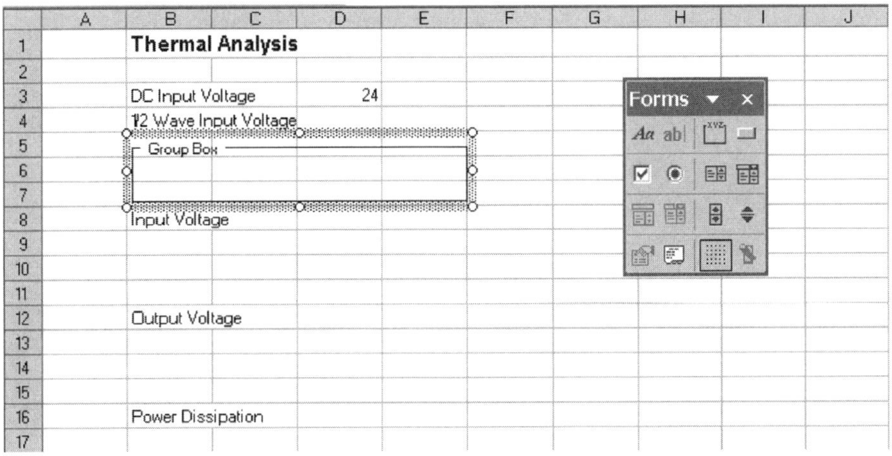

Figure 10-14: Creating a Group box.

Click on a cell away from the Group box and then move the cursor over the text of the group box until the cursor becomes a four-headed arrow. Then right-click and select **Edit Text** from the pop-up menu. Change the Group box title to something like "Input Voltage Selection   " with a few spaces at the end to improve the appearance.

Click on the Option button in the toolbox. Click within the Group box, and drag a window (within the Group box) to a suitable size. There are to be two Option buttons in this box, but rather than creating a second, select the first by right-clicking on it and cutting and copying. There is another way to copy a control. **<Ctrl> + <Click>** on the original control and then drag while still holding the **<Ctrl>** key.

By copying the control, both buttons and associated text will be the same size. Right-click on each and modify the text. Also, right-click on either one and select the **Format Control** and point the cell link to cell A7. If you want to copy this setup (and I certainly will), it is better to make this a relative and not an absolute reference. Click away from the button to lose the focus, and then clicking on one or other of the buttons will change the value of cell A7 from 1 to 2 and back. See Figure 10-15.

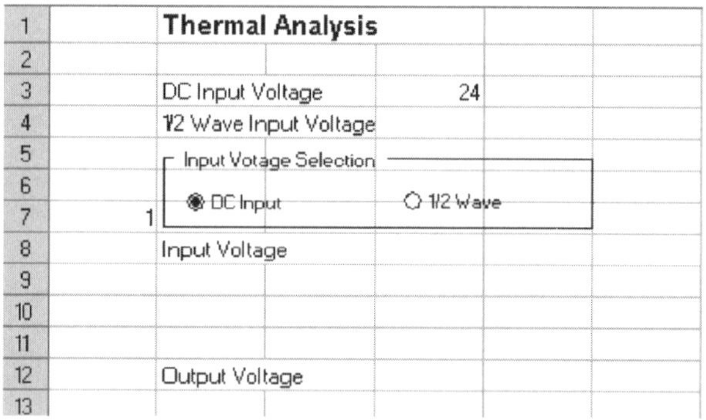

Figure 10-15: Using Option buttons.

1		**Thermal Analysis**	
2			
3		DC Input Voltage	24
4		1/2 Wave Input Voltage	
5		Input Votage Selection	
6		● DC Input ○ 1/2 Wave	
7	1		
8		Input Voltage	
9			
10			
11			
12		Output Voltage	
13			

We now add other options, either by copying and pasting or by starting fresh each time until we arrive at Figure 10-16.

There are several types of component packages, but for simplicity I have stayed with the TO-220. There are different methods of affixing the LM317 to the heat sink. In the one that uses the Kapton insulator (Sil-Pad®), the thermal conductivity varies with the pressure affixing the component to the heat sink. I have stayed with one value.

Figure 10-17 shows the formulas used in the calculations. The user is expected to enter the current through the device in cell D15 and the maximum ambient temperature in cell D16. The power dissipation in the device is found using equation (3), as previously shown.

The overall thermal conductivity is derived from equation (1), and if the result is greater than 50 °C/W (derived from the data sheet entry "Thermal Resistance, Junction-to-Ambient (No Heat Sink)") then no heat sink will be required and this will be annunciated in cell

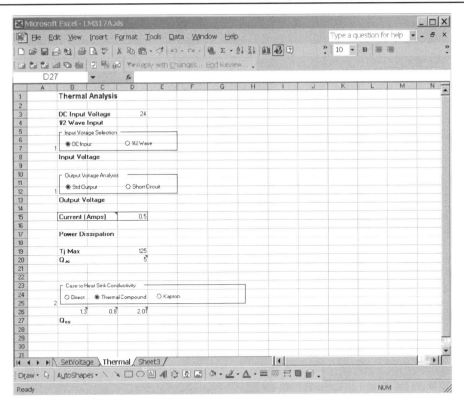

Figure 10-16: Preparatory work on options and inputs.

E32. Otherwise, the required thermal resistance is calculated from equation (1) and (2) and reported in cell D32. Figure 10-18 shows the completed thermal model.

In Parenthesis: *CHOOSE*

The format of the CHOOSE function is:

CHOOSE(index_num,value1,value2,...)

index_num selects which entry in the following list is used. It can be a numerical value, evaluate to a numerical value or refer to a cell containing a numerical value, but it cannot exceed the value of 29. In other words, the maximum size of the following list is 29.

The following list can be a value, a calculation, or refer to a cell. It can even refer to an array of cells. For instance:

=min(choose(1,b7:b25,b26:b30,b31:b45))

will effectively reduce to:

=min(b7:b25)

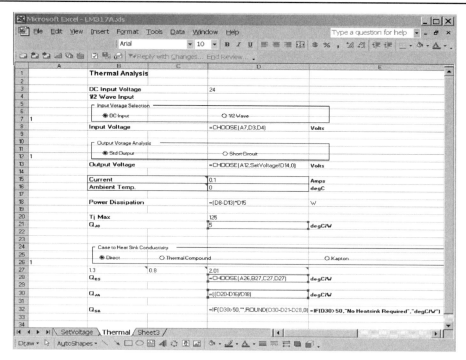

Figure 10-17: Formulas needed to calculate the thermal conductivity for a heat sink.

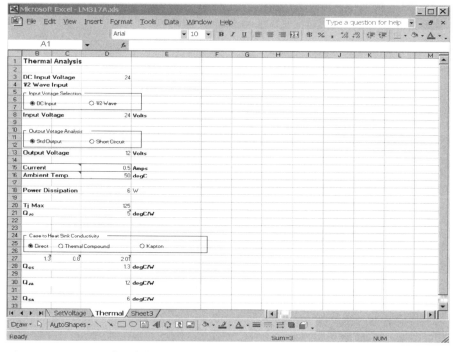

Figure 10-18: Completed thermal analysis.

Half-Wave Rectification

It is very common to provide a rectified and smoothed voltage as a source to a voltage regulator. Throughout the building automation sector, 24VAC is used as a supply voltage with one of the sides tied to chassis ground. The simplest way of converting this to a DC voltage (with a ripple on it) is through half-wave rectification as shown in Figure 10-19.

Figure 10-19: Half-wave rectification circuit.

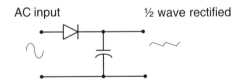

The minimum value for the input voltage to the voltage regulator must not drop below the dropout voltage of the regulator, so a large smoothing capacitor would reduce the ripple. On the other hand, the less the value of the smoothing capacitor, the smaller and cheaper it is likely to be. In addition, the effective voltage (RMS voltage) is reduced and consequently the power dissipated is also reduced, economizing on the requirements for the heat sinking of the regulator. We can use Excel to calculate the optimal value of this capacitor.

True RMS and Integration

Part of the model will calculate the RMS value of the voltage to use in the calculation of the power dissipated in the regulator. Before we examine the complex waveform of the smoothed half-rectified AC, let us test the model as to how we are going to calculate a finite integral using Excel and we will use a sine wave since we know what the results should be. The RMS voltage is calculated from the equation:

$$V_{rms} = \sqrt{\frac{1}{T}\int_0^T v^2(t)\,dt}$$

Integration between limits defines area under a curve, so by dividing the area into trapezoids, we can calculate the area of each trapezoid and sum them to calculate the total area. The area of a trapezoid is the average of the sum of the two parallel sides multiplied by the distance between them. This is shown in Figure 10-20. The area of one of the trapezoids is $((Y1 + Y2)/2) * X1$. Obviously, the smaller the value of X1, the greater the accuracy of the calculation.

Figure 10-20: Trapezium method of calculating the area under a curve.

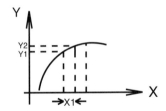

If we take a formula for a curve and evaluate it for a number of points, we can use these points as the values for Y1 and Y2 and so calculate the area.

Figure 10-21 shows the formulas used to implement this for a sine wave. I have hidden some of the middle of the range points (rows 14 to 43) to fit the top and bottom of the worksheet into the figure. I have chosen to work with 50 Hz since the numbers are nicer, and anyway when we get to the smoothing capacitor, the result will be that the design can be used in the rest of the world as well as North America. The formula for a sine wave is $A_0 \sin(2\pi ft)$ where A_0 is the peak amplitude, f is the frequency of the wave, and t is the elapsed time. The 0.001 factor that appears is the conversion of milliseconds to seconds.

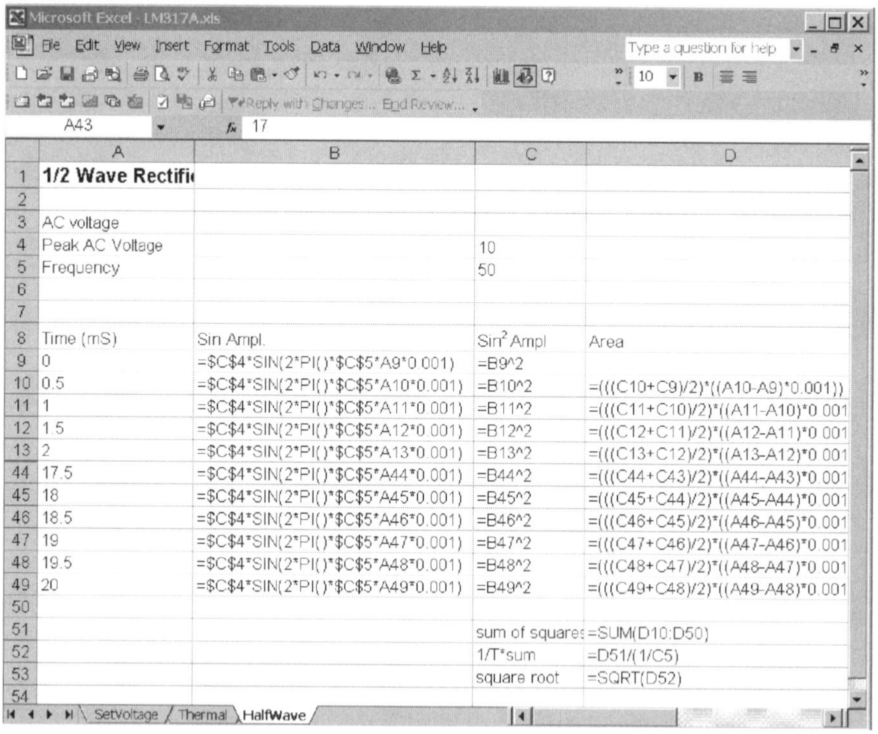

	A	B	C	D
1	**1/2 Wave Rectifi**			
2				
3	AC voltage			
4	Peak AC Voltage		10	
5	Frequency		50	
6				
7				
8	Time (mS)	Sin Ampl.	Sin^2 Ampl	Area
9	0	=$C$4*SIN(2*PI()*$C$5*A9*0.001)	=B9^2	
10	0.5	=$C$4*SIN(2*PI()*$C$5*A10*0.001)	=B10^2	=(((C10+C9)/2)*((A10-A9)*0.001))
11	1	=$C$4*SIN(2*PI()*$C$5*A11*0.001)	=B11^2	=(((C11+C10)/2)*((A11-A10)*0.001
12	1.5	=$C$4*SIN(2*PI()*$C$5*A12*0.001)	=B12^2	=(((C12+C11)/2)*((A12-A11)*0.001
13	2	=$C$4*SIN(2*PI()*$C$5*A13*0.001)	=B13^2	=(((C13+C12)/2)*((A13-A12)*0.001
44	17.5	=$C$4*SIN(2*PI()*$C$5*A44*0.001)	=B44^2	=(((C44+C43)/2)*((A44-A43)*0.001
45	18	=$C$4*SIN(2*PI()*$C$5*A45*0.001)	=B45^2	=(((C45+C44)/2)*((A45-A44)*0.001
46	18.5	=$C$4*SIN(2*PI()*$C$5*A46*0.001)	=B46^2	=(((C46+C45)/2)*((A46-A45)*0.001
47	19	=$C$4*SIN(2*PI()*$C$5*A47*0.001)	=B47^2	=(((C47+C46)/2)*((A47-A46)*0.001
48	19.5	=$C$4*SIN(2*PI()*$C$5*A48*0.001)	=B48^2	=(((C48+C47)/2)*((A48-A47)*0.001
49	20	=$C$4*SIN(2*PI()*$C$5*A49*0.001)	=B49^2	=(((C49+C48)/2)*((A49-A48)*0.001
50				
51			sum of squares	=SUM(D10:D50)
52			1/T*sum	=D51/(1/C5)
53			square root	=SQRT(D52)
54				

Figure 10-21: Formulas to calculate the RMS value of a sine wave. Note that the worksheet has been renamed.

Cells B9 to B49 calculate the amplitude of the sine function at different times. Note the use of the *PI( )* function for the π value. Cells C9 to C49 contain the square of the amplitudes. Cells D10 to D49 contain the calculation for the area of each trapezoid, and the areas are all summed in D51. This value is divided by the period (*1/f*) in cell D52, and the square root is found in cell D53. The results are shown in the worksheet in Figure 10-22 (please excuse the lack of formatting), and the result is very close to reality. See the actual results in Figure 10-22.

Now that we have considered the calculation of the RMS voltage, we can put it aside for a while. I have left this workbook as *LM317_Sine.xls.*

	A	B	C	D	E
1	**1/2 Wave Rectification**				
2					
3	AC voltage				
4	Peak AC Voltage		10		
5	Frequency		50		
6					
7					
8	Time (mS)	Sin Ampl.	Sin² Ampl	Area	
9	0	0	0		
10	0.5	1.56434465	2.447174185	0.000611794	
11	1	3.090169944	9.549150281	0.002999081	
12	1.5	4.539904997	20.61073739	0.007539972	
13	2	5.877852523	34.54915028	0.013789972	
44	17.5	-7.071067812	50	0.028862712	
45	18	-5.877852523	34.54915028	0.021137288	
46	18.5	-4.539904997	20.61073739	0.013789972	
47	19	-3.090169944	9.549150281	0.007539972	
48	19.5	-1.56434465	2.447174185	0.002999081	
49	20	6.43149E-15	4.1364E-29	0.000611794	
50					
51			sum of squares	1	
52			1/T*sum	50	
53			square root	7.071067812	
54					

◄ ◄ ► ►◄ \ SetVoltage / Thermal \ HalfWave / ◄

Figure 10-22: Calculation of the RMS value of a 50 Hz sine wave with an amplitude of 10.

More Preparation

In a half-wave rectifier, the smoothing capacitor is charged until the AC voltage peaks. It then discharges according to the formula $i = C dv/dt$ where i is the current, C is the capacitance of the capacitor, dv is the change in voltage and dt is the change in time. The AC voltage continues to drop until it reaches zero and stays zero until the next positive cycle starts. In the meantime, the capacitor discharges linearly (since the current through the regulator is constant) until the increasing AC voltage exceeds the reduced capacitor voltage whereupon the capacitor is recharged.

Actually, the capacitor discharge may not start exactly at the AC peak, but it should be close enough for this calculation.

I have created the top of the worksheet to include all the parameters that are needed (see Figure 10-24), and the cells C3 to C9 have been suitably named. Only the nominal transformer voltage is required as an entry from the user. This whole effort is to find the capacitance, but to initiate the development an arbitrary value is entered. All the other cells are derived.

Let's create the table of the AC waveform. We will start the analysis from the time 5 mS since this is where the AC signal peaks and the capacitor starts to discharge, and continue it to 25 mS, which is where the AC signal next peaks. Each cell with the amplitude calculation (B10 to B50) contains the following formula (adjusted for relative cell locations):

=IF(ac*SIN(2*PI()*freq*A10*0.001)>=0,ac*SIN(2*PI()*freq*A10*0.001),0)

to allow for the fact that the signal is at 0 in the negative half of the sine wave as a result of the rectification.

The voltage drop *dv* is given by:

dv=i*dt/C

and this is calculated in cells C13 to C53. The entry is:

=current*(A13–$A$13)*0.001/(Cap*0.000001)

The capacitance is converted to farads by the factor 0.000001 in the denominator.

D13 to D53 have the resulting droop generated by subtracting *dv* from the peak voltage that the capacitor was charged to:

=ac–C13

We now combine the two voltages in cells E13 to E53. The higher of the two voltages becomes dominant by use of the following formula:

=IF((D13>B13),D13,B13)

This traces the waveform as it charges and discharges the capacitor.

The MIN function in Excel simply looks at a range of numbers and returns the minimum value. Cell E55 contains the formula:

=MIN(E13:E53)

which is the minimum value of the regulator input voltage. We would like this minimum voltage to be no lower than the regulated output voltage plus the dropout voltage of the regulator. From the data sheet, we pick a safe dropout voltage of 2.5V, entered as a constant in cell C8.

Now we use the Goal Seek tool. It will be set up to change the value of the capacitor (cell c5="Cap"), while monitoring the cell E55 (the minimum voltage) for the value of the dropout voltage plus the output voltage.

In order to do this, we follow the sequence **Tools | Goal Seek** and the dialog window pops up as in Figure 10-23.

Right away we notice a problem that is hinted to by the lack of the expand button on the right-hand side of the **To value:** entry. Excel requires a number here, it cannot handle a cell reference. This is easy enough to solve by recording this Goal Seek process as a macro. The result of this, recorded to the macro named *FindCapacitance* in the example, follows:

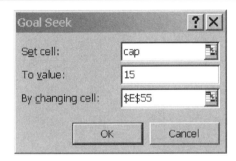

Figure 10-23: Using Goal
Seek to determine the
capacitor value.

```
Sub FindCapacitance()
    Range("E55").GoalSeek Goal:= 15, ChangingCell:=Range("C6")
End Sub
```

We edit the macro to read:

```
Sub FindCapacitance()
    Range("E55").GoalSeek Goal:=(Range("Vreg").Value + Range("Vdrop").Value), ChangingCell:
        =Range("C6")
End Sub
```

and this will automate the process. Figure 10-24 shows the progress so far.

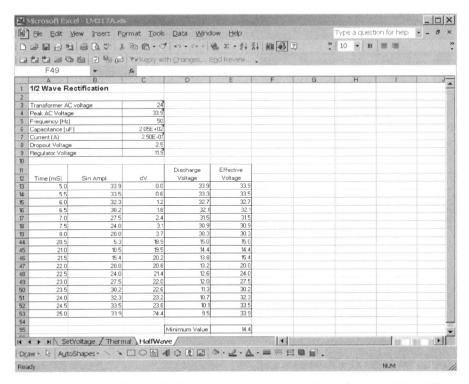

Figure 10-24: Calculation of minimum capacitance value. Note that rows 20 to 43 are
hidden.

Standard Capacitance Value

I am sure that it comes as no surprise to you that as with resistors, there are standard capacitance values as well. Since the smoothing capacitors are only likely to be between 10 μF and 10000 μF, there are very few values to consider, so I have just created a new worksheet and entered the possible values in a vertical column (Figure 10-25).

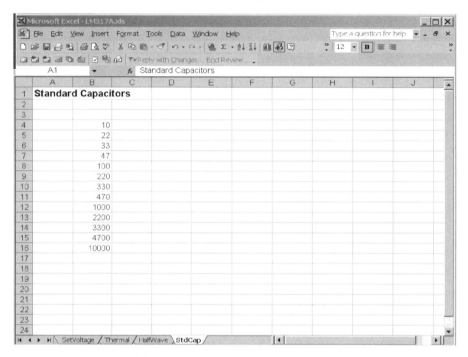

Figure 10-25: New worksheet with standard capacitor values. To add a worksheet, right-click on the sheet tabs and click on the Add Worksheet icon.

If we enter a formula in cell E6 of the HalfWave sheet:

=vlookup (Cap,StdCap!B4:B16,1)

the value returned is the entry below the desired capacitance. In this instance, we want the capacitor value greater than the calculated value, so we first need to fetch the identified location using the MATCH function to find the associated row, and then use the INDEX function to get the next value up. Cell E6 becomes:

=INDEX(StdCap!B4:B16,(MATCH(Cap,StdCap!B4:B16)+1),1)

Having calculated the standard capacitance, we need to reevaluate the input waveform. Cells F13 to F53 contain the formula (suitably transposed for relative cells):

=ac-current*(A13-$A$13)*0.001/(StdCap*0.000001)

which represents the discharge curve of the capacitor, and cells G13 to G53 contain the resultant waveform when combined with the AC input. As before, while the capacitor's decaying voltage is higher than the AC input, it is the dominant voltage. Once the AC input exceeds it, it becomes dominant and the capacitor recharges. The formula is:

=IF((F13>B13),F13,B13)

This column forms the basis for the RMS value calculation. Column H contains the square of the input voltage (for example, G13^2), and then using the Trapezium method as detailed earlier, each trapezoid area is calculated in cells I14 to I53 using the formula:

=(((H14+H13)/2)*((A14-A13)*0.001))

Note that there is no entry for cell I13 as there is no previous value to use. From this point, it is easy to add each calculated area segment to get the total area under the curve (in cell I55) and to multiply by the inverse of the period (cell I56) and then take the square root (cell I57) for the RMS voltage. This is the number that we should use in the **Thermal** worksheet for non-DC inputs (if you remember we had deferred that issue). So on the **Thermal** worksheet, cell D4 becomes:

=Vrms

Note that a named cell does not need to have a sheet reference with it.

Figure 10-26 has the results of this calculation. In terms of the sequence of data entry, it seems to me that the **HalfWave** worksheet should be before the **Thermal** worksheet. It is

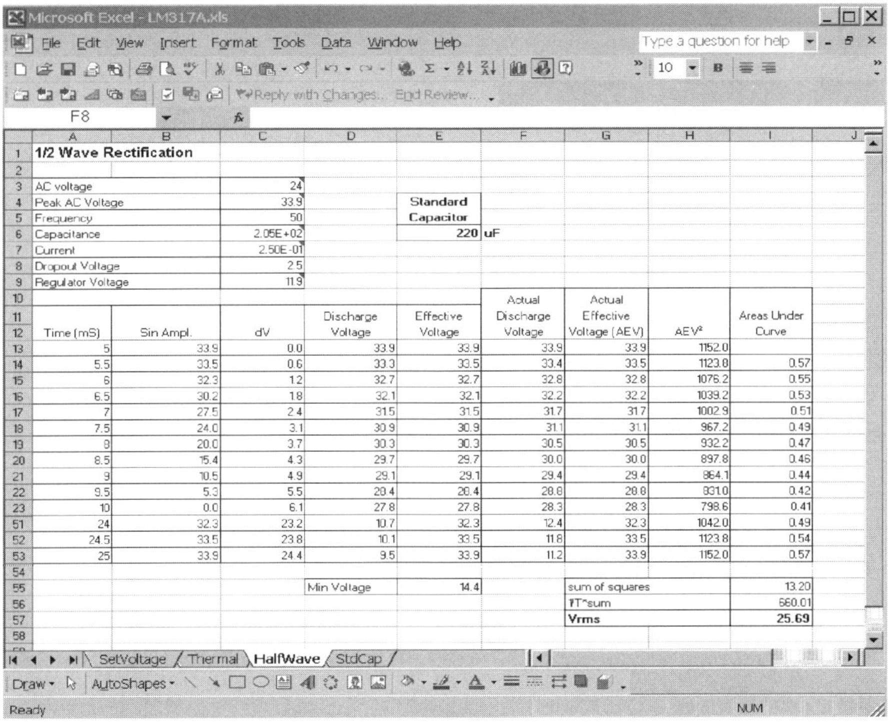

Figure 10-26: Completed worksheet—almost!

easy enough to rearrange. Click the **HalfWave** tab, then drag the tab to the left of the **Thermal** tab until a small black triangle pops up just above the insertion point and then release the mouse button.

I added a Command button that triggers the FindCapacitance macro at the top of the worksheet.

Chart

It would be nice to have a graphical representation of the ripple, so let's introduce a chart. With the **HalfWave** sheet selected, select cells A13 to A53 and G13 to G53. Click on the Chart icon on the standard toolbar, or follow the toolbars **Insert** | **Chart** and select the **Standard Type** tab. Select the options as shown in Figure 10-27 and click on **Next**.

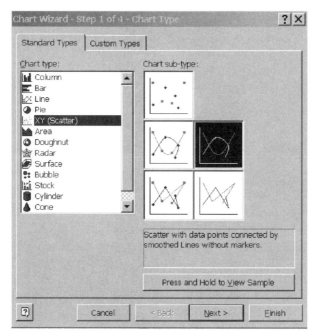

Figure 10-27: Creating a chart.

Having preselected the ranges, we do not need to modify anything in step 2, although sometimes Excel does not correctly interpret your desires. Click on **Next**. We are now given the opportunity to add some cosmetic effects to the chart. We can add titles to the axes, a chart title, gridlines and more (see Figure 10-27). Once more, click on **Next** to get to the fourth step.

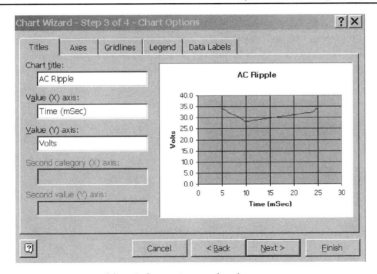

Figure 10-28: Adding information to the chart.

The final step allows us to place the chart on the sheet or elsewhere. I preferred to place it on the same sheet with the result in Figure 10-29.

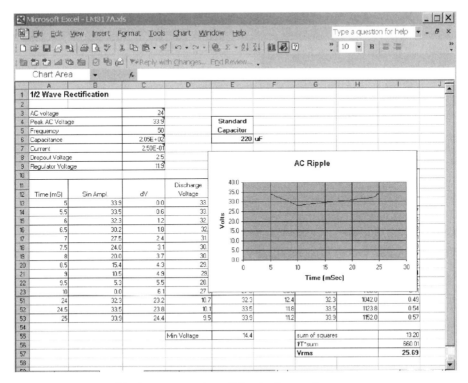

Figure 10-29: Graphical representation of the ripple waveform.

This shows an interesting effect in Excel. You will notice an irregularity in the 20 to 25 mSec area and it doesn't seem to get anywhere near the expected minimum. If we expand the hidden cells, this is the chart that we get (Figure 10-30). That's more like it! This effect can be turned off in the **Tools** | **Options** | **Chart** sequence. It can also be used to your advantage on a chart with a large number of entries, using every fifth reading, say.

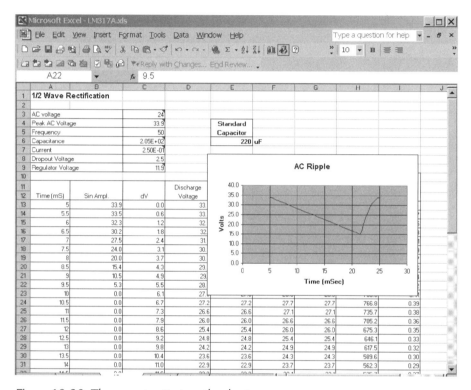

Figure 10-30: The correct output on the chart.

Right-clicking on almost any aspect of the chart will allow you to change the object's properties. For instance, you can change the number of "ticks," and the font and alignment on an axis. Go ahead and try a few!

With all its versatility, the chart model apparently doesn't allow you to add a freehand line, which I would like to add to indicate the absolute minimum, the line y=14.4 in our particular case. The simplest way to do this is to enter =$E$55 in cell J13 and copy it to cells J14 to J53. Then right-click on the chart and select **Source Data,** or go through **Chart** | **Source data.** Click on the **Series** tab, and then click on the **Add** button (see Figure 10-31). Define the new series and click on **OK**. This will have the desired effect with the result in Figure 10-32. Another shortcoming of the Chart utility is that it is not possible to add random text, and as a result, if you want to identify which line is which, you need to name each series and enable the **Series name** option under the **Data labels** tab in the **Chart Options** dialog.

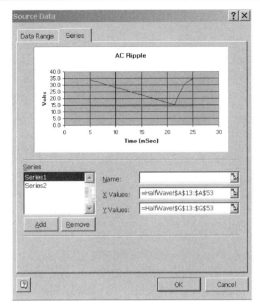

Figure 10-31: Adding a new series to a chart.

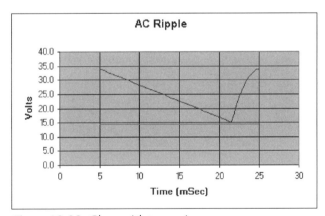

Figure 10-32: Chart with two series.

Conclusion

This has been quite a broad area to cover as a single model, and as a result I have tried to keep it simple. I have not included all the possible tolerances on the components that could have an effect on the outcome. The tolerance of the capacitor for instance could be ±20%.

Development of the model through stages has lead to some inconsistencies in data entry and data flow. For instance, irrespective of whether the thermal or half wave analysis is done first, data is needed from one to feed the other. The model would benefit from adding the DC current to the **SetVoltage** worksheet, possibly with an input box in the LM317 macro. Although the model could use a little polish it does show quite how useful Excel can be.

EXAMPLE **11**

TL431 Adjustable Voltage Reference

Model Description

Like the LM317 in the previous example, the TL431 adjustable voltage reference has permeated throughout the industry. Its simple configuration, low cost and wide adjustment ability are the features that have endeared it to electronics engineers. The basic schematic is shown in Figure 11-1.

Figure 11-1: Connections to the TL431.

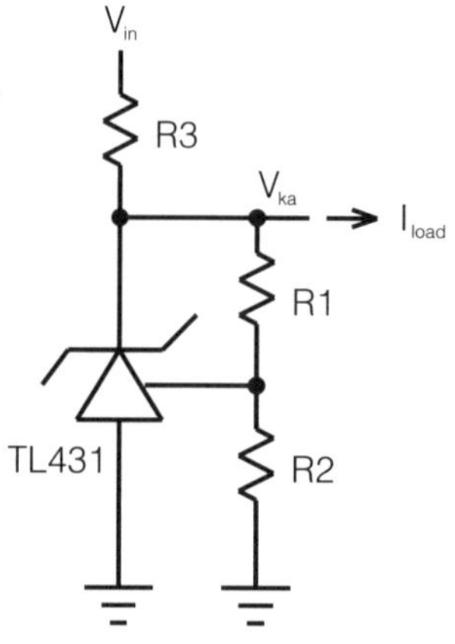

The output reference voltage V_{ka} is defined by the ratio of the two resistors R1 and R2, the device reference voltage V_{ref} and the current flowing into the reference terminal of the device I_{ref}. The formula is:

$$V_{ka} = V_{ref}(1 + R1/R2) + I_{ref} * R1$$

Installing the NearestValues Add-In

If you have not installed the NearestValues add-in, follow the instructions in the section titled *Installing the NearestValues Add-In* in Example 9. This function will allow us to look up standard resistor values.

Initial Model

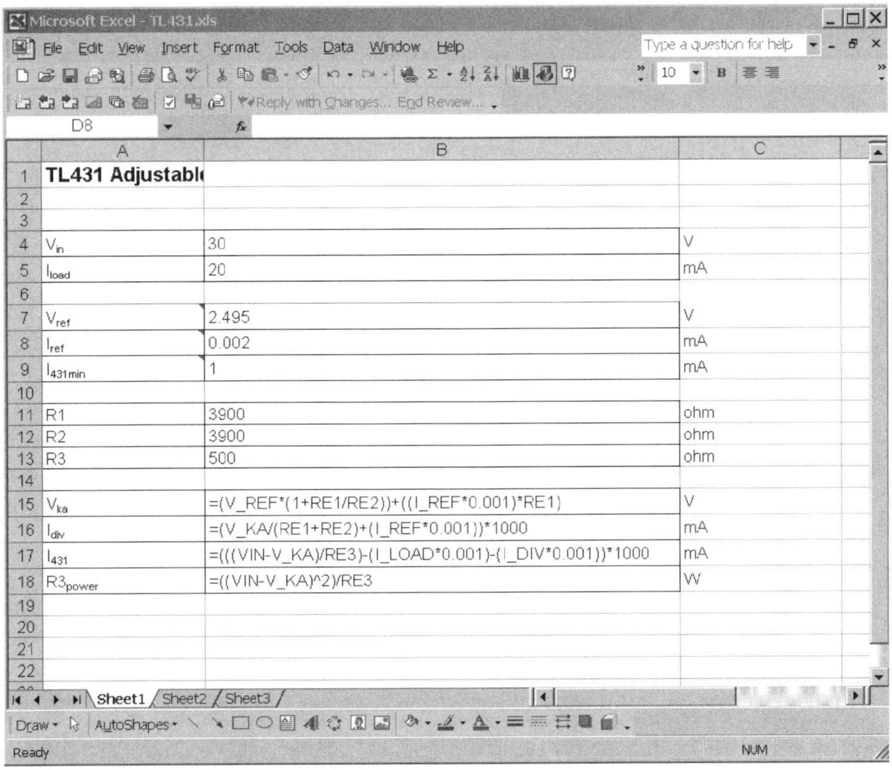

	A	B	C
1	TL431 Adjustabl		
2			
3			
4	V_{in}	30	V
5	I_{load}	20	mA
6			
7	V_{ref}	2.495	V
8	I_{ref}	0.002	mA
9	I_{431min}	1	mA
10			
11	R1	3900	ohm
12	R2	3900	ohm
13	R3	500	ohm
14			
15	V_{ka}	=(V_REF*(1+RE1/RE2))+((I_REF*0.001)*RE1)	V
16	I_{div}	=(V_KA/(RE1+RE2)+(I_REF*0.001))*1000	mA
17	I_{431}	=(((VIN-V_KA)/RE3)-(I_LOAD*0.001)-(I_DIV*0.001))*1000	mA
18	$R3_{power}$	=((VIN-V_KA)^2)/RE3	W

Figure 11-2: Initial setup.

In doing the analysis, aside from the output voltage, there are other constraints that need to be evaluated. For reasonable results, we must know what the system requirements are: system supply voltage (V_{in}), and the current to the load (I_{load}). The TL431 needs at least 1 mA through it (I_{431min}) to guarantee that it regulates correctly. The resistive divider of R1 and R2 also loads the regulator output voltage (V_{ka}) and we would like this to be as small as possible (I_{div}). Finally, we would like R3 to be as small as possible so as not to limit I_{load}, yet large enough to prevent excessive power dissipation in it.

Figure 11-2 shows the initial setup of the formulas for this evaluation. R1, R2 and R3 have arbitrary values for the moment to check out the model. Note the factors of 0.001 and 1000 within some of the formulas, which are required for milliamp to amp conversions and back. The Excel file on the CD-ROM is called *TL431.xls*.

Solver

In earlier examples, we have seen that Goal Seek can change the value of a cell while monitoring the result in another cell, stopping when the target cell reaches a chosen value. Goal Seek works well when we can reduce the problem to a single variable. If you remember, in the LM317 example we set the value of one of the resistors to that recommended in the data sheet. Solver is the tool we reach for when there are two or more variables that we need resolved.

In setting up, we let Solver know what cells to change and which cell to monitor as an output. In this aspect, it is very similar to Goal Seek, but it also allows us to apply the constraints that I mentioned in the last section. To invoke Solver, click on **Tools | Solver**. Up pops the dialog window in Figure 11-3. Add the information into the entry boxes as shown.

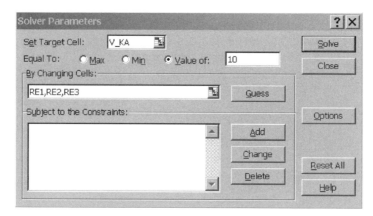

Figure 11-3: Defining the cells to change, and the result cell to monitor.

Note that aside from looking for an exact value in the target cell, we can also look for a minimum or a maximum. The cells that may be changed can be entered as an array (block) or individually using the **<Ctrl> + <Click>** for multiple cells. You can run the Solver now by clicking on the **Solve** button. It will attempt to solve the problem, but the minute it finds a solution it will stop—there is no guarantee that the solution meets our requirements or even reality! For instance, a negative value resistor may work nicely in theory.

First, let's set up the global constraints. Click on the Options button in the Solver Parameters dialog (Figure 11-3). A new dialog box appears with a whole bunch of arcane options (translation: I am having trouble understanding some of them), as shown in Figure 11-4. We will discuss some of them later. For the moment, just check the **Assume Non-Negative** option and then click on the **OK**. This will force Solver to only consider positive numbers for the inputs.

You will be returned to the dialog of Figure 11-3. Click on the **Add** button and enter the data in the dialog as in Figure 11-5. Click on **Add** for a new constraint, and add constraints as in Figure 11-6 and Figure 11-7. Click on **OK** to return to Figure 11-8, which provides a summary of these settings. Note the use of cell names, which have been previously defined.

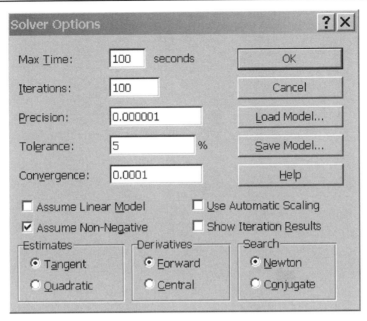

Figure 11-4: Forcing all input values to be positive or zero.

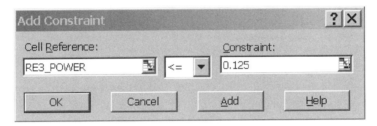

Figure 11-5: Limiting the power dissipation in R3 to 1/8W (assuming a 1206 size resistor).

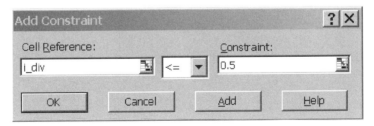

Figure 11-6: Limiting the current through the resistor combination R1 and R2.

Figure 11-7: Ensuring that there is sufficient current through the TL431.

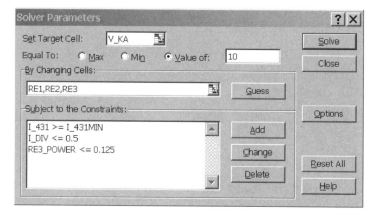

Figure 11-8: Solver parameters dialog showing the constraints.

Now click on **Solve** and … it failed to find the answer. Since I am trying to make a point here, you didn't really expect success at the first try, did you? See Figure 11-9.

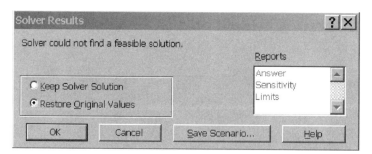

Figure 11-9: Reevaluation needed! Click on Restore Original Values and then OK.

Programs that work on an iterative basis like Solver (and even P-SPICE), take certain values and try to adjust them in certain ways and see how the relationships respond. The conditions that Solver starts from are based on the numbers that we enter as the seed—the initial numbers in the worksheet. It certainly speeds up the calculation, and may even allow convergence, to have a good idea of what the answer is likely to be and use that as the seed. The actual algorithm that Solver uses to generate a solution is hidden from us and we cannot influence its predictability, but we can step through each iteration to try to get an idea why the solution is unattainable.

In the Solver Options dialog (Figure 11-4), check the **Show Iteration Results** option, and run the Solver. It will pause after each iteration to allow an evaluation of the perturbations to the input cells and the results on the outputs. The final result is shown in Figure 11-10.

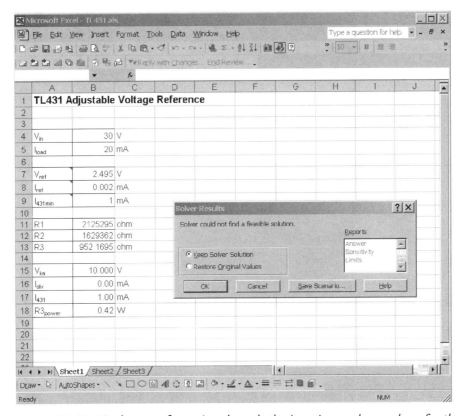

Figure 11-10: Final stage of stepping through the iterations—the numbers for the variables will depend on the initial values.

As we stepped through, you may have noticed that cell B18 ($R3_{power}$) never dropped to anywhere near 0.125W, so we should consider increasing this. Click on **Restore Original Values** and then **OK**.

Go to the Solver Parameters dialog, click on the constraint for R3 power and then click on **Change**. Modify the constraint to **<=0.5**. Also, go to the Solver Options and disable **Show Iteration Results** option. Now try to solve and it works! Click on the **Save Scenario** button and name it *FirstAttempt*.

Return to the Solver Results dialog, and click on OK, so that the new results remain on the worksheet. Modify the values of R1, R2, and R3 to 200. Now using these values, run Solver again (**Tools | Solver,** and click on **Solve**). Name this scenario as *SecondAttempt.*

We can now view the scenarios by clicking on **Tools | Scenarios** and in the Scenario Manager (Figure 11-11) viewing each scenario. Note that the results are different, yet both are an acceptable solution.

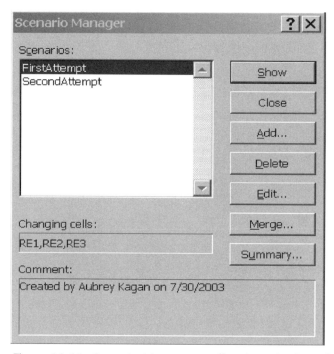

Figure 11-11: Scenario Manager to allow investigation of different conditions.

As we play around with the Solver settings trying to find a solution, it is possible to save each group of settings by clicking on the **Save Model** button in the Solver options dialog. These settings are saved to a series of cells in the worksheet after the user is prompted for the location that the user wants them to be stored at. It stands to reason that it is possible to reload the settings with the **Load Model** button.

In Parenthesis: *Solver Options*

If we look at the different Solver Options (Figure 11-4), some are obvious as to their impact. The maximum time limits the amount of time that the solver will attempt to find a solution. The number of iterations will cap the number of times it tries to solve the problem. Exceeding either condition will terminate the attempt.

Precision affects what Solver will accept as a value in a constraint cell. It is a number between 0 and 1. The closer it is to 0, the closer the resulting value will be to the target value. Conversely, the closer it is to 1, the less precise the result. One of the options on setting a constraint is to set a number to an integer or a binary number. However, having an exact number can cause Solver to expend excessive time on its way to a solution. The mathematical definition of an integer can be relaxed by adjusting the precision.

The Tolerance setting pertains to the target cell and is only valid if there are constraints restricting the inputs to integers. It represents the percentage within which the result will be considered acceptable.

Convergence is only valid for nonlinear models (**Assume Linear Model** option is un-checked). If the value of the target cell changes by less than the convergence value in the last five iterations, the solver process is terminated. The smaller the number, the longer it takes for Solver to reach a solution. The convergence value can be between 0 and 1. Obviously, if every part of the model is linear, setting the **Assume Linear Model** option will result in quicker results.

The **Show Iteration Results** option forces solver to pause after iteration and is used in analyzing and changing the model to lead to a solution.

If inputs and outputs have large difference in magnitudes (a small number in the de-nominator of an expression can lead to very large answers), select the **Use Automatic Scaling** option.

The Estimates, Derivatives and Search settings determine how Solver approaches find-ing a solution. If you genuinely need to know about these, then you are well beyond the scope of this book and I will be looking to you for advice.

The Estimates selection determines the way the initial guesses are made for the input cells. Advice I have seen recommends **tangent** for linear problems and **quadratic** for nonlinear.

The Derivatives selection chooses the differencing method for partial derivative estimates. In other words, when successive changes in the input cells results in a slow result change, the **Forward** option should be used (for most problems). For a very sensitive response, use the **Central** option.

The choice in Search affects the algorithm used in determining the size of memory needed for the calculations. The **Newton** selection uses more memory, but because it requires less iterations is quicker. The **Conjugate** selection uses less memory, but takes longer. It may be preferred in larger models where memory usage is an issue.

Some more detailed information can be found in the Microsoft Knowledge Base numbers 82890 and 214115.

The truth is, for simple models the defaults are normally more than sufficient. If you are having a problem, then a few well considered constraints can work wonders.

In Parenthesis: *Use of Constraints*

Constraint Relationships:

There are five forms of constraints that can be set: >=, =, <=, integer and binary. The Solver evaluates the constraint and decides if it is valid within certain limits. These limits are fixed by the Precision setting in the Solver Options. For instance, if one of the constraints was =B33=0, then if the cell value was 0.001 (given the nature of computer calculation), the strict evaluation of the constraint would result in a false condition. Setting the precision to a number closer to 0 makes the acceptance of the value more stringent. As I understand it, a Precision of 0.00001 would not accept the above cell value as equal zero, but 0.002 would.

A similar philosophy applies to the integer constraint and the binary constraint. In true mathematical terms, an integer is 3, not 3., or 3.0 or 3.000001. In the Solver world, however, it may be acceptable and as before, the precision defines the acceptability. Extrapolating on this concept: True is 1 and False is 0 within the Precision limits, and these could be used as a go/no go decision.

Cell Reference (left-hand side of constraint):

The entry here can be any individual selection of cells. They can be a single cell or a single block of cells (column, row, or rectangle), but they cannot have multiple selections.

Constraint Entries (right-hand side of constraint):

Frontline Systems (creators of Solver) recommend that the constraint entry (right-hand side) is a constant or links to a cell (or cells) that contains a constant. The cell reference can be a single cell or a group of cells (for example, B3:B42) or even a formula like C2*B4. Of course, cell names can be used. If you have more than one cell (as in the B3:B42 example), the number of cells must match the number of cells on the left-hand side. Multiple cell selections are permitted and they do not have to have the same cell pattern on the worksheet, but they will be matched element for element, that is, the third item on the left-hand side will be evaluated to the third item on the right-hand side.

Standard Resistor Values

Let's add the feature of looking up standard resistor values. Provided we have enabled the add-in as described at the beginning of this example, we can insert the function (if we don't remember its name) as follows: click on cell D11, and then **Insert** | **Function** and the dialog of Figure 11-12 appears.

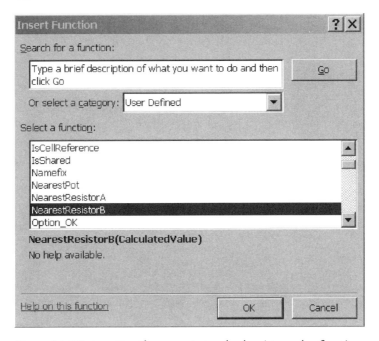

Figure 11-12: Inserting the nearest standard resistor value function.

Select **User Defined** and then **NearestResistorB** options, and for the parameter enter RE1 and **OK**. Repeat or copy the formula to cells D12 and D13 editing the parameter to RE2 and RE3, respectively. We then name the cell D11 as *RE1_S*, and so on.

In cell D15, we enter the formula:

=(V_REF*(1+RE1_S/RE2_S))+((I_REF*0.001)*RE1_S)

which will reevaluate the voltage output in terms of the standard resistors. After some cosmetic adjustment, Figure 11-13 is the result.

Adding a Macro

For multiple usage of the model, a macro would probably be a good idea. A word of caution here: there are two ways to record the macro, with or without the constraints. The constraints are maintained, so there is no real need to include them in the macro and it will allow the user to change them without having them reset when running the macro. On the

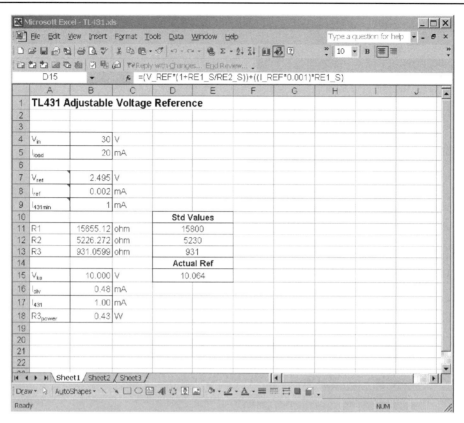

Figure 11-13: Worksheet with standard values added.

other hand, once changed, they will stay modified the next time the application is run. Of course it is possible to add a second macro to set the constraints back to the default. I am opting not to change the constraints within the macro.

As before, we start the process by learning the Solver sequence. **Tools | Macro | Record new macro**, and name it *Solve*. Then click on **Tools | Solver | Solve**. Leave the update worksheet option **Keep Solver Solution** (although this will not show up in the macro, and you will always be given this option). The macro is recorded as:

```
Sub Solve()
    SolverOk SetCell:="$B$15", MaxMinVal:=3, ValueOf:="10", ByChange:= _
      "$B$11,$B$12,$B$13"
    SolverSolve
End Sub
```

The first time you run this macro, however, will likely end up with an error message. This is because although you will have added the Solver as an add-in in Excel, it has not been added in VBA. The way to solve this is to get to the VBA editor (if it is not already open: **Tools | Macro | Visual Basic Editor** is one method). In VBA, click on **Tools | References,** and in the **Available References** window click on **Solver.xla**. If it does not exist in the list, click on Browse and find it on your hard drive. It is normally in the **Office\Library** subfolder. Mine was in:

C:*Program Files\Microsoft Office\Office 10\Library\Solver*

Now the macro will run without detecting an error. Let's create a Command button on the worksheet so that running the macro repeatedly will not take too many key clicks. Using **View | Toolbars | Forms,** click on the Command button icon in the Forms toolbox and then click and drag a button on the worksheet. Associate the Command button with the *Solve* macro and change the button text to "Solve".

If we added a few user prompts for the input voltage, desired reference voltage and load current, it would surely streamline the process so let's modify the macro as follows (note the *<space>_* line continuation):

```
Sub Solve()
    Dim nValueOf

    Range("RE1").Value = 100
    Range("RE2").Value = 100
    Range("RE3").Value = 100
    ' intitiate from the same point
    Range("Vin").Value = (InputBox("Enter input voltage", _
            "TL431 Configuration", "24"))
    nValueOf = (InputBox("Enter reference voltage", _
            "TL431 Configuration", "5"))

    Range("I_load").Value = Val(InputBox("Enter desired load current", _
            "TL431 Configuration", "1"))

    SolverOk SetCell:="$B$15", MaxMinVal:=3, ValueOf:="nValueOf", ByChange:= _
        "$B$11,$B$12,$B$13"
    SolverSolve
End Sub
```

I also added an initiation of the resistor values to add some consistency to the results. The input boxes sequentially prompt the user for data, and then save it in the worksheet, or pass it through to the solver function. The appearance of one of the input boxes is shown in Figure 11-14.

Figure 11-14: Input box. Note the prompt, title and default values as related to the function call.

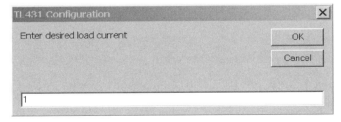

Report Generation

At the final stage of the Solver function when a solution has been found, you must have noticed the option of a report as in Figure 11-15.

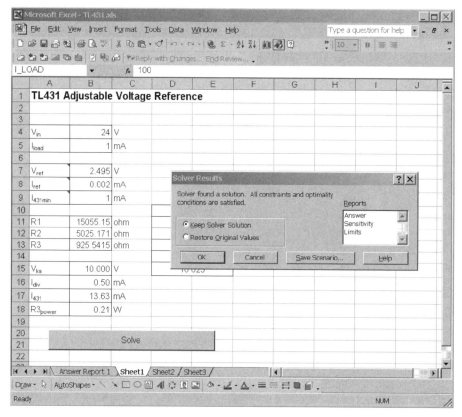

Figure 11-15: Successful conclusion of a Solve process.

These reports can be used to help analyze the solution. Clicking on any one of the Report options will result in a new sheet being added to the workbook with the report. Figure 11-16 shows the **Answer Report,** which is largely a summary of the solver action. Figure 11-17 indicates the **Sensitivity** of the data to change, while Figure 11-18 indicates the **Limits** placed on the changes that took place. These are of much more significance in complex models.

Within the reports:

"Slack" is the difference between the actual solution and the constraint target value.

"Reduced Gradient" shows how a unit increase in a cell value will affect the solution.

"Legrange Multipliers" indicates the amount by which the solution will improve as a result of a relaxation of the constraint by one unit.

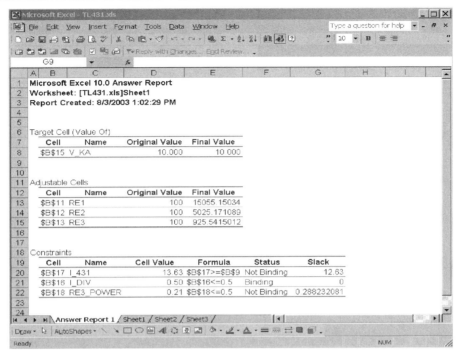

Figure 11-16: Answer report worksheet.

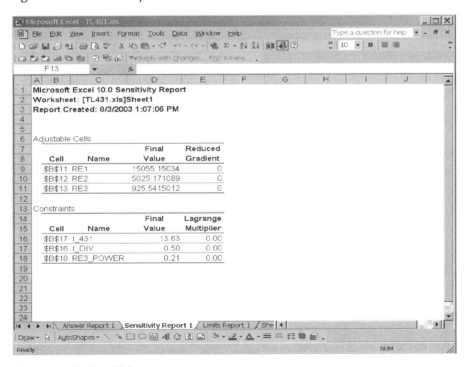

Figure 11-17: Sensitivity report.

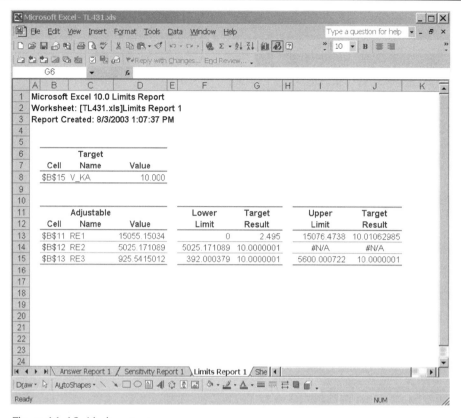

Figure 11-18: Limits report.

Limitations

It occurred to me that using Solver to find resistor and capacitor values and then fetching the standard value may result in solutions that may be bettered by starting out using the standard resistor and capacitor values. I set out to find a method to achieve this. My solution was to create a table of standard values (as opposed to the NearestValue approach). I wanted to access the standard value by using an INDEX function based on an integer that would vary. I hope Figure 11-19 gives you the idea of what I was trying to achieve. It is on the CD-ROM as *Index.xls*.

Unfortunately, Solver only applies the "integerization" of cell A4 once it approaches a solution. It also requires a change in the target cell, no matter how small, within 5 iterations or it terminates, unable to find a solution. In this case, it sets A4 to 1.0000001 as the first step. Since the INDEX function returns the exact same value, you might think that Excel should try a larger value in cell A4. It does not—it simply maintains the value. It seems to me that Excel must see some change in the target cell in order to calculate what to try next since the value in A4 is not modified in the next iteration. (Ironically, the equivalent feature in Quattro Pro, called *Optimize*, does allow for this and works well. Optimize was also developed

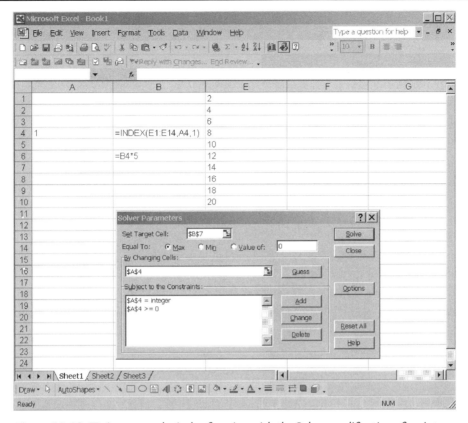

Figure 11-19: Trying to use the index function with the Solver modification of an integer in A4. E2 to E20 contain general data, just to prove the point.

by Frontline Systems, the developer of Solver in Excel.) Microsoft has confirmed to me that Solver will not work with a lookup function. There are upgraded versions of Solver available from Frontline Systems (www.solver.com), but I don't know whether they can handle this approach either.

This discussion serves as a segue to the next example, where I will use the Solver as part of a procedure to partially achieve this operation.

Incidentally, Excel includes an example called *Solvsamp.xls*, which should be on your hard disk. Included on one of the worksheets is an example of the application of Solver to an RLC circuit.

EXAMPLE 12

555 Timer

Model Description

The 555 timer and its descendants still enjoy great popularity in the electronics world. This "bubble-gum" part owes its longevity to simplicity of use, versatility of configuration, robust operation and economical pricing. Hardly suitable for an Excel model you might think, but do not be hasty; it will still provide a challenge for some of Excel's abilities. Hopefully, we will learn something along the way.

The principle configurations of the 555 are a monostable or an astable multivibrator (Is multi-vibrator an archaic term? Is there a newer word for it? Am I showing my age?). There are many other functions that may be realized, but the treatment is very similar and so they are left to you (as my lecturers used to say, as an exercise). I will deal with each function independently, and they will be implemented on two separate worksheets of the workbook *555Timer.xls*.

Monostable Operation

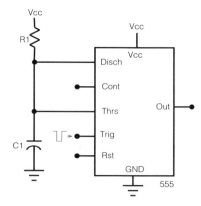

Figure 12-1: 555 timer as a monostable multivibrator.

Figure 12-1 shows the connections of the 555 when configured for monostable operation. When a trigger occurs (input voltage less than Vcc/3), the output of the 555 goes high for a period of:

$T = 1.1 * R1 * C1$

Provided the trigger has returned to a level above Vcc/3 in the interim, the output returns low and the system is ready for the next trigger.

Setup

Let us create a new workbook setup as in Figure 12-2. Note that Sheet1 has been renamed as *Monostable*. Cell C8 contains the formula:

$$=1.1*CAP1*1E-12*RE1*1E-6$$

where the numerical factors are used for scaling the capacitors to farads and the time to microseconds.

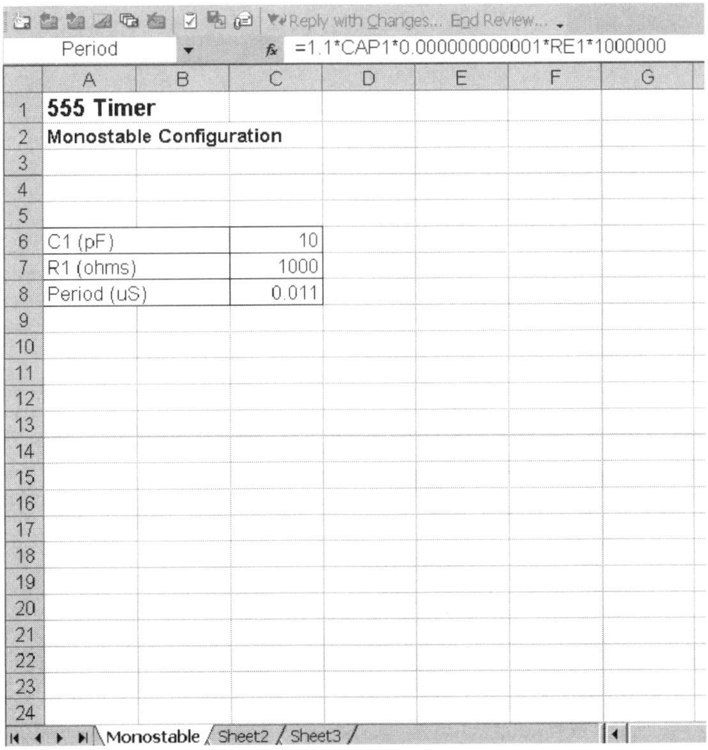

	Period	▼	f_x	=1.1*CAP1*0.000000000001*RE1*1000000			
	A	B	C	D	E	F	G
1	**555 Timer**						
2	**Monostable Configuration**						
3							
4							
5							
6	C1 (pF)		10				
7	R1 (ohms)		1000				
8	Period (uS)		0.011				
9							
10							
11							
12							
13							
14							
15							
16							
17							
18							
19							
20							
21							
22							
23							
24							

Monostable / Sheet2 / Sheet3 /

Figure 12-2: Preparation for monostable model.

Add User Form

Cosmetically, it would be a nice touch to add the schematic of Figure 12-1 right in the worksheet. Invoke the VBA editor (**Tools** | **Macros** | **Visual Basic Editor** or <Alt> + <F11>). Detailed explanations of the setup used here are covered in Example 6. In the editor, click in **Insert** | **User Form** and modify the forms properties by clicking in the right-hand column of the associated property:

(**name**): *frmMonostable*

Caption: *Monostable Configuration* (title at the top of the form).

Height: *200* (fixed height to keep consistency with the other image).

Width: *260* (fixed width to keep consistency with the other image).

StartUpPosition: *1- Center Owner* (will appear in the middle of the Excel window).

ShowModal: *False* (to allow the form to be shown an still enter data).

BackColor: *White* (selected from palette—white to blend in with image that we will add shortly).

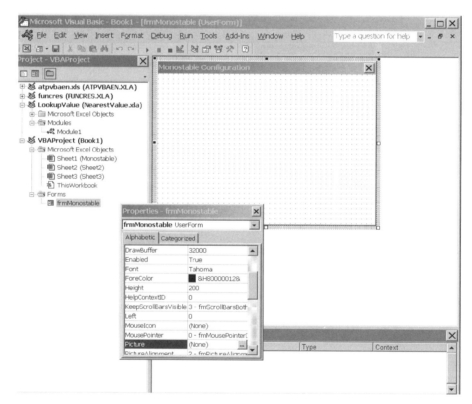

Figure 12-3: Adding a user form.

Add Image Control

We now want to add the image into the user form. We need the VBA Controls toolbox. If it is not visible, click on **View | Toolbox**.

Click on the Image Control icon in the toolbox, and then click and drag on the user form to define an area.

Modify the Image box properties as follows:

Picture: *monostable.wmf* (find on disk. As mentioned in Example 7, the wmf format seems to give me better results than any other format.)

BackColor: *select white on the palette* (to blend in with the user form).

BorderStyle: *0-fmBorderStyleNone* (no border).

Autosize: *True* (image automatically sizes to window).

Height: *100* (shrinks image to size that can be handled within user frame).

Width: *100* (shrinks image to size that can be handled within user frame).

Size and position the image by dragging the "handles" on the edges so that it fills the user form and looks right.

Figure 12-4 will give you an idea of the result.

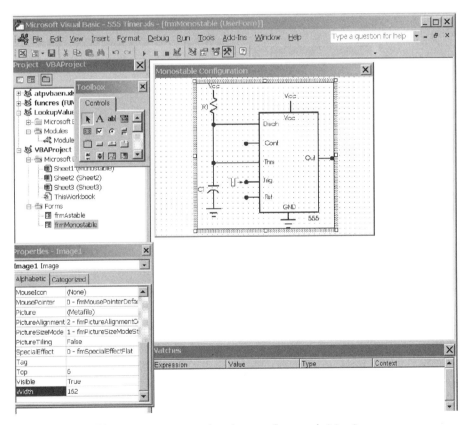

Figure 12-4: Adding an image control to the user form and sizing it.

Second Image

Let's repeat the process for a second diagram for the astable configuration, even though we are getting a little ahead of ourselves. We create form frmAstable and insert an image control using the image astable.wmf.

While we are so far ahead, let's also rename Sheet2 to *Astable*.

Turning Forms On (and Off)

We will need a procedure to turn all the forms off, since we will do this from several places. Double-click on the **ThisWorkbook** folder in the 555 Timer VBA project, right-click on it and select **Insert Module**. Double-click on the newly added **Module1** and then **Insert | Procedure**. Name the procedure *HideForms*, and select the options as in Figure 12-5.

Figure 12-5: Adding a procedure.

Add the code so that the procedure appears as follows:

```
Public Sub HideForms()
'hide all forms
    frmAstable.Hide
    frmMonostable.Hide
End Sub
```

Next, double-click on the **Sheet 1(Monostable)** folder in the VBA Explorer, and click on the drop-down button of the Object bar, selecting the Worksheet option as in Figure 12-6. Then click on the drop-down button of the Procedure bar and select Activate. Enter the code as shown in Figure 12-6.

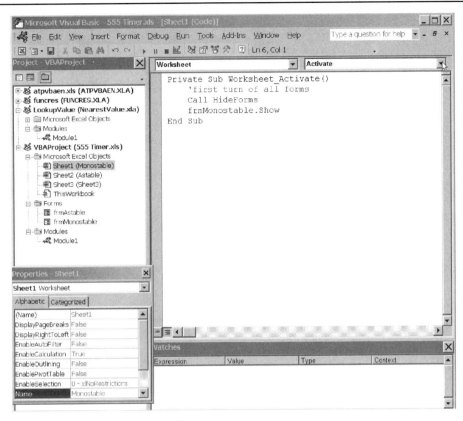

Figure 12-6: Selecting and modifying sheet events.

We also need to do the same process for the second sheet. Double-click on Sheet2(Astable) and find the associated Worksheet_Activate event. Add the code as follows:

```
Private Sub Worksheet_Activate()
    'first turn of all forms
    Call HideForms
    frmAstable.Show
End Sub
```

Try clicking between sheets, and the two different diagrams should pop up in the respective sheet.

We also need to initialize the workbook. Double-click in **ThisWorkbook** folder, and select the **WorkbookOpen** event. Add this code to close all forms and then activate the first sheet (and as a consequence showing the monostable diagram).

```
Private Sub Workbook_Open()
    Call HideForms
    Sheets("Monostable").Select
End Sub
```

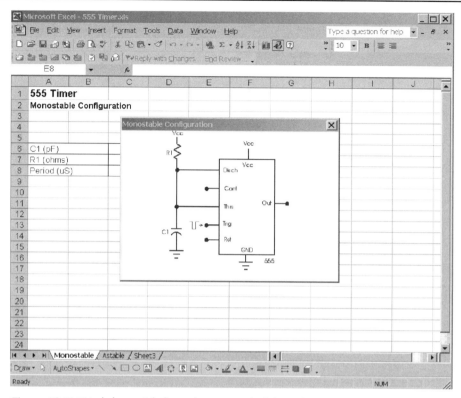

Figure 12-7: Worksheet with form shown smack-dab in the middle of the window.

Modifying Form Location

As can be seen from Figure 12-7, placing the picture in the middle of the worksheet may prove irritating in operation, so let's change the process to place the form in the lower right-hand of the window. First, we must change the **StartUpPosition** property of both forms to **0 – Manual**. This will prevent the position of the form from being reinitialized every time the Show procedure is executed. Then we modify both event procedures as follows:

For the Monostable sheet activation:

```
Private Sub Worksheet_Activate()
    'first turn of all forms
    Call HideForms
    'positioning the form at the bottom
    frmMonostable.Top = Excel.Application.Top _
        + Excel.Application.Height _
        - frmMonostable.Height
    'positioning the form at the right
    frmMonostable.Left = Excel.Application.Left _
        + Excel.Application.Width _
        - frmMonostable.Width
```

```
        frmMonostable.Show
    End Sub
```

and for the Astable sheet activation:

```
    Private Sub Worksheet_Activate()
        'first turn of all forms
        Call HideForms
        'positioning the form at the bottom
        frmAstable.Top = Excel.Application.Top _
            + Excel.Application.Height _
            - frmAstable.Height
            'positioning the form at the right
        frmAstable.Left = Excel.Application.Left _
            + Excel.Application.Width _
            - frmAstable.Width

        frmAstable.Show
    End Sub
```

In Parenthesis: *More on Combo Boxes*

In all the examples to date, all the controls that we have placed on a worksheet have been drawn from the Forms toolbox. These controls are simple to use, but have several disadvantages. First, they cannot be turned on and off, so they cannot be made to simply disappear when they are not needed and then reappear. Second, when there are several of the controls, it is not always easy to set them up to the same size or on the same horizontal or vertical line. For better or worse, the control "floats" above the worksheet and the selection must be referred to a cell on the worksheet.

It is possible to get a drop-down effect right in a cell, so that the number is embedded and directly accessible. To do this, select a cell and click on **Data | Validation**. In the window that pops up (Figure 12-8) under the **Settings** tab, you can define what kind of data entry that will be accepted (decimal, text, and so forth), and the upper and lower limits, where applicable. If a list is chosen, then a series of cells can be used for the input or a list separated by commas can be used. Under the **Input Message** tab, you can create a message that will show when the cell is selected (Figure 12-9) and you can provide an error message if the function rejects the input value under the **Error Alert** tab (Figure 12-10).

Aside from the Combo box available in the Forms toolbox, there is also a Combo box option available in the Control Toolbox. Find the Control Toolbox by clicking on **View | Toolbars | Control Toolbox**. These are the controls with Excel and not part of VBA.

The Combo box control has a nice feature in that when the user starts to enter data, the selection that is displayed is refined as the user types. Unfortunately, it does not validate the input. The in-cell approach does. It is possible to combine the two approaches by applying in-cell validation to the output cell of the Combo box. For the data validation entries, select **List** and make sure the **In-cell** drop-down is unchecked.

Monostable Pulse Width Entry

In previous examples, I set up the macro to prompt for the target value using an Input box. Let's take a new approach.

Click on cell B5. Click on **Data | Validation**, and fill in the data as in Figure 12-8, Figure 12-9 and Figure 12-10.

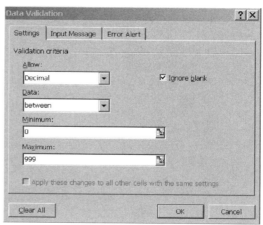

Figure 12-8: Allowing any decimal number between 0 and 999.

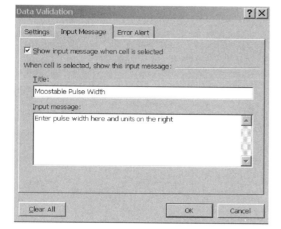

Figure 12-9: Message shows when cell is selected.

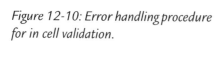

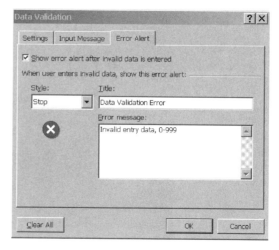

Figure 12-10: Error handling procedure for in cell validation.

If we try to enter an erroneous value in B5, Figure 12-11 is the result.

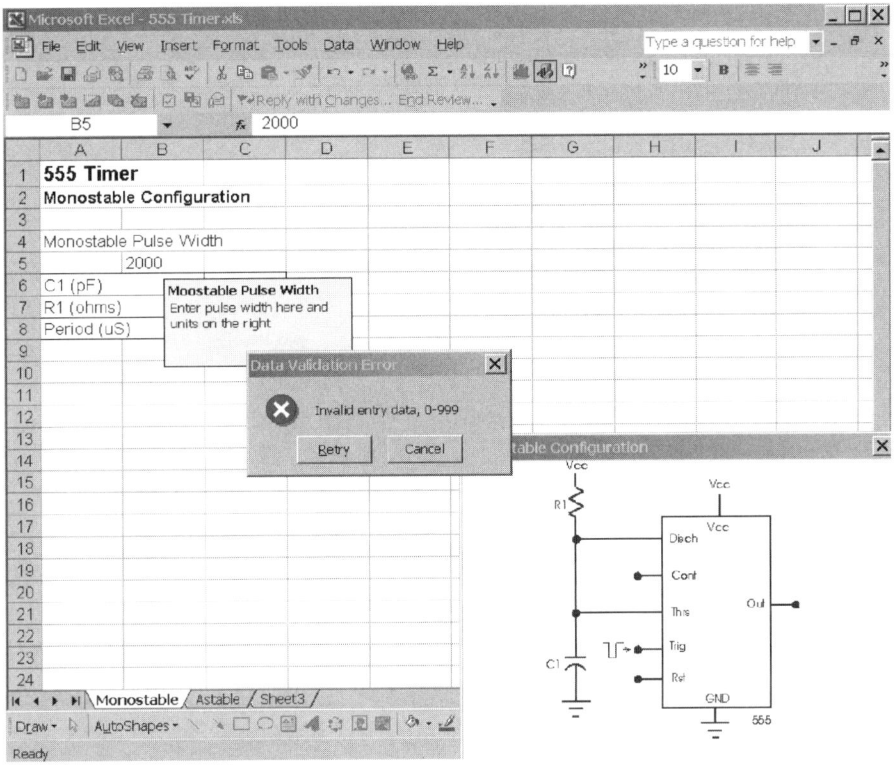

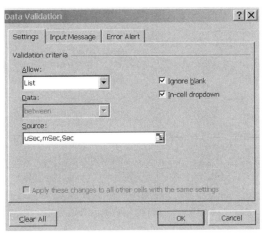

Figure 12-11: In-cell validation. Note the message from the cell selection and the error message for erroneous data.

In addition, let's set up cell C5 to allow for units of microseconds, milliseconds and seconds, as follows in Figure 12-12.

Figure 12-12: List data for the units of the pulse width.

It is possible to click in C5 and enter data, but it will be rejected if it is not the same as the list. You will notice that when the cell is selected, the drop-down arrow appears to the right of the cell and this can be used to enter the desired units.

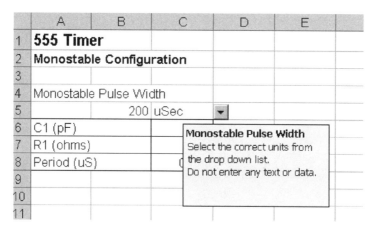

Figure 12-13: Entry of pulse width time units.

We now need to make a single number out of these two entries. In cell E5, we enter the formula:

$$=IF(C5="uSec",1,IF(C5="mSec",2,IF(C5="Sec",3,0)))$$

which selects a number between 0 and 3 depending on the units selected.

Cell E6 has the formula:

$$=IF(AND(E5<>0, B5<>0),CHOOSE(E5,B5,B5*1000,B5*1000000),0)$$

Provided that there are valid entries, this formula uses the lookup function CHOOSE to return the valid calculation of the desired pulse width time in microseconds.

Command Button

Running ahead of ourselves for the second time in this example, we will be using a button to execute the macro (which is still to be written). Rather than use the Forms control button, which cannot be hidden or enabled, we are going to use the Control Toolbox Command button. When there is a valid entry for the time, the button will be enabled, otherwise it will be disabled and the macro cannot be run. That is the plan anyway!

Place the Command button on the worksheet by working through **View** | **Toolbars** | **Control Toolbox**. Click on the Command button icon, and then click on the worksheet and drag out an area for the button as in Figure 12-14. Right-click on the button and select **Properties**. In the properties window, edit the name to *cmdSolve* and the caption to *Solve*.

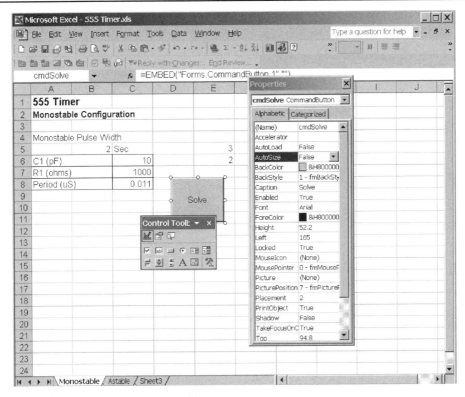

Figure 12-14: Placing a command button.

After much trial and error and investigation, it seems to me that there are some limitations on enabling and disabling a control (see "In Parenthesis: Control Toolbox"). The most elegant solution I found was to place the following code in the worksheet change event. Every time anything on the worksheet changes, this code is run.

```
Private Sub Worksheet_Change(ByVal Target As Range)
    If Range("e6").Value <> 0 Then
        Sheet1.cmdSolve.Enabled = True
    Else
        Sheet1.cmdSolve.Enabled = False
    End If
End Sub
```

In Parenthesis: *Control Toolbox*

As a result of historical development, I guess, Excel has two forms of "in-sheet" controls. Up until now, we have only considered the Forms controls. They are much easier to use, but lack versatility. If I were to prognosticate, I might say that these controls would gradually fade out in favor of the controls in the Control Toolbox.

Using the Control Toolbox will produce controls that act much as the controls from the Forms toolbox, but they are much more like VBA objects and their properties are available without having to go into VBA. This allows them to be sized, enabled, and made visible and invisible. Like the Forms controls, it is possible to link to a cell which has a value associated with the control output. This cell is accessed through a property of the control and obviously accessed through the properties window.

Unlike the Forms controls, you cannot associate just any VBA procedure with these controls. The procedure must be invoked from the events associated with the controls. Of course, these events can call any procedure you want.

Inserting or editing the controls requires that the Control Toolbox be in Design Mode, which is attained when the Design Mode button is pressed in. When the button is out, Excel is in Run Mode, and the effects of the controls can be seen.

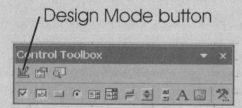

Figure 15: Control Toolbox toolbar.

As usual, right-clicking on the object (in Design Mode) brings up the options for the object. Note that sometimes you will need to click away from the object and then right-click on it to get the correct options to show up.

Having said that, it is possible to enable or disable or make visible controls from the Control Toolbox. I should mention that from my observations, this is only possible from within certain events. While it is possible to modify properties (like the **caption**) from a VBA function or any event, the **Enabled** and **Visible** properties appear to be changeable only from events that are triggered by a click. In most of the other events, although the code is executed, the property does not change. I have not found any documentation on this, so it is based on trial and error. In this example, I have found that these properties can also be changed in the Worksheet_Change event.

It appears that it is not possible to change any of the object properties directly from a worksheet formula. Of course it can be accessed through a VBA function contained in the worksheet, subject to the above limitation.

Solver

Returning from the tangent that we went off on, we want to find the component pair R1 and C1 for a given pulse width. For this, we resort to Solver. If we plunge headfirst into it, the result will be values for R1 and C1, without any regard to the fact that they can only have certain specific values. Now this is not normally a problem for resistors since there are many values, but the standard range of values for capacitors has some sizeable gaps and the result may not be reasonable.

As we saw at the end of the last example, it is not possible to use lookup techniques within Solver, so what I will present is a method of fixing a capacitor in VBA and then calling Solver. If the result is unacceptable, the next value of capacitor in a list is selected and so on through the range of capacitors until a result is found.

First, let's construct a simple model to show that everything works. We will not use the cell for C1 as an input, allowing Solver just to find the value for R1. See Figure 15.

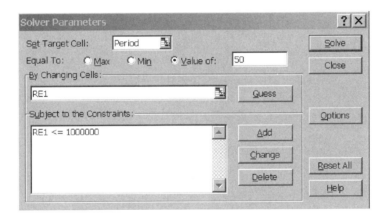

Figure 12-16: First attempt at Solver. In Options, the Assume Non-Negative is selected. R1 is limited to 1 MOhm.

The target value of 50 μS is arbitrary at the moment, since it is not possible to directly link this to a cell.

Now we record each step of this to a macro including a **Reset All** to start off with. This is the macro that should result:

```
Sub Solve()
'

    SolverReset
    SolverOk SetCell:="$C$8", MaxMinVal:=3, ValueOf:="50", ByChange:="$C$7"
    SolverAdd CellRef:="$C$7", Relation:=1, FormulaText:="1000000"
    SolverOk SetCell:="$C$8", MaxMinVal:=3, ValueOf:="50", ByChange:="$C$7"
    SolverOptions MaxTime:=100, Iterations:=100, Precision:=0.000001, AssumeLinear _
        :=False, StepThru:=False, Estimates:=1, Derivatives:=1, SearchOption:=1, _
        IntTolerance:=5, Scaling:=False, Convergence:=0.0001, AssumeNonNeg:=True
```

```
    SolverOk SetCell:="$C$8", MaxMinVal:=3, ValueOf:="50", ByChange:="$C$7"
    SolverSolve
End Sub
```

Notice that the **ValueOf** property is assigned at each step. This is the value that we want to associate with a cell, but we only really need to change it in the last instance. That line becomes:

```
    SolverOk SetCell:="$C$8", MaxMinVal:=3, ValueOf:=Range("$e$6").Value, ByChange:="$C$7"
```

In addition, I also added a limitation that R1 should be greater than 100 ohms. This restriction looks like:

```
    SolverAdd CellRef:="$C$7", Relation:=3, FormulaText:="100"
```

In the cmdSolve_Click event we add the line:

```
    call Solve
```

Any time we click the button, the Solver is run for whatever time period is entered. However, this only modifies the value of R1. We now need to find a way to introduce standard capacitor values.

Standard Values

There are times when the approach used to find standard component values in the Nearest-Values functions is not suitable. To remedy this, I have created a worksheet with the standard values of resistors and capacitors. It will be easy enough to cut and paste this to any workbook, so I am not going to attempt to create any form of standard interface as I have with other functions. You will find it in the "555 Timer.xls" workbook on the StandardValues sheet.

Before you think that this is a lot of work, I would like to point out that with a little thought it is not that much. We only need to enter a range of data for the first decade and then create the second decade as 10 times greater, and so forth. For instance, if cells C5 to C19 contained the standard capacitor values, then cell C10 would have the formula = 10* C5. This is then copied through the entire remainder of the range. Some care should be taken since the first decade(s) is (are) missing some values. It is actually easier to create the second decade and then work from there.

Once all the values have been created, highlight the whole range (using click and drag or similar) and copy it using **<Ctrl> + <C>** or **Edit | Copy**. Then go through **Edit | Paste Special,** and select **Paste Values** and **OK**. All the formulas are simply transformed into values.

SolverSolve

The function call that does all the work on the solver is called *SolverSolve*. There does not appear to be much documentation out there on the subject, so I have had to improvise and find my way through trial and error. This is the third time in two examples that I have bumped into the limits of Excel. Is this book cutting edge or what? (To those of you that lean toward the "what" side of the equation, please note that I am being self-deprecating.)

One of the things that you may have noticed when clicking on the **Solve** button is that the Solver always reports the result of the calculation, but the information does not appear to be available to the calling procedure. This is simply resolved by looking for the return value from SolverSolve. The last line of the macro is changed to:

varSolveReturn = SolverSolve(True)

where varSolveReturn is defined as a variant at the start of the procedure. We should also note that the Solver completion message is suppressed in the function call.

In Parenthesis: *SolverSolve Function*

As a function call, SolverSolve employs the format:

SolverSolve(UserFinish, ShowRef)

UserFinish is a Boolean value of True or False. If it is True, then no Solver results are displayed. False or nothing will lead to the solver results being displayed.

ShowRef is a macro name that will be executed at each Solver iteration if the **Show Iteration Results** option is set.

Undocumented are the return values. Empirically, I have found that 0 corresponds to success and the value of 4 represents a failure to find a value. There are other conditions that Solver detects like division by 0 and other errors, but on the assumption that the model is first perfected using the normal procedures, I have not tried to find the other possibilities. Microsoft could not give me the values and suggested that I use their "Pro" support (read: services that an individual cannot afford). They could not tell me if in fact they did have this information.

In Parenthesis: *Calling an Excel Function from VBA*

While it is possible to create any function you want in VBA, sometimes it is quicker and easier to use an existing Excel function. The magic words are:

Application.WorksheetFunction.

Casting this incantation at the beginning of the function will allow access to the function, provided there is no function within VBA that has the same name. VBA will even provide the prompts that you would expect to see from Excel.

Using Standard Capacitor Values

The process I intend to follow is to start with a low capacitor value (we will set the minimum value to be 100 pF) and try to solve using Solver. If a solution is found, then the process terminates. Otherwise, the next standard capacitor value is fetched and a new attempt made for a solution. If there are no more capacitor values, an error message is generated, but instead of creating our own, we will user the solver message. The philosophy in this approach is to find the lowest value capacitor that will achieve the result. The modified procedure Solve looks like this:

```
Sub Solve()
'

    Dim varSolveReturn As Variant
    Dim intl As Integer

    For intl = 16 To 63
    'starting at 100pF and limiting at 10uF
        Range("c6").Value = Application.WorksheetFunction.Index(Worksheets("Standardvalues").
Range("D5:D82"), intl)
        SolverReset
        SolverOk SetCell:="$C$8", MaxMinVal:=3, ValueOf:="50", ByChange:="$C$7"
        SolverAdd CellRef:="$C$7", Relation:=1, FormulaText:="1000000"
        SolverAdd CellRef:="$C$7", Relation:=3, FormulaText:="100"
        SolverOk SetCell:="$C$8", MaxMinVal:=3, ValueOf:="50", ByChange:="$C$7"
        SolverOptions MaxTime:=100, Iterations:=100, Precision:=0.000001, AssumeLinear _
          :=False, StepThru:=False, Estimates:=1, Derivatives:=1, SearchOption:=1, _
          IntTolerance:=5, Scaling:=False, Convergence:=0.0001, AssumeNonNeg:=True
        SolverOk SetCell:="$C$8", MaxMinVal:=3, ValueOf:=Range("$e$6").Value, ByChange:="$C$7"
        varSolveReturn = solversolve(True)
        If varSolveReturn = 0 Then
            Exit For
            'force a break
        End If
    Next intl

    If varSolveReturn <> 0 Then
        solversolve
        'rerun with failed values and no update of C1.
    End If
End Sub
```

If you run this, you will actually see every update and it is quite slow. We can speed it up by preventing the screen update from occurring. If we add:

Application.ScreenUpdating=False before the For statement and the inverse just before the End Sub statement, it does improve the situation a little.

Tidying Up

Let's just add a summary at the bottom with the Standard values found with the result in Figure 12-17.

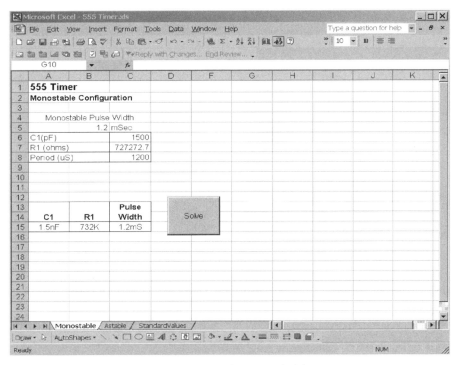

Figure 12-17: Finished worksheet. Note column E is hidden.

Cells A15 to C15 have some formatting to allow expression of the numbers in a more "normal" format. For instance, B15 has the formula:

$$=IF(\$E\$7<1000,\$E\$7\&"R",(IF(\$E\$7<1000000,\$E\$7/1000\&"K",\$E\$7/1000000\&"M")))$$

Unfortunately, Solver is incompatible with any level of protection on the worksheet, so we are left with the possibility of inadvertently changing some of the cells.

Astable Operation

Figure 12-18 shows the schematic of the astable configuration for the 555 timer. The frequency of oscillation is:

$\dfrac{1.44}{(R1+2R2)C1}$, and the duty cycle is calculated from $\dfrac{R2}{R1+2R2}$. The duty cycle can only vary from 0% to 50%. Depending on how you view the duty cycle, some texts look at the inverse of the signal and the duty cycle would then vary from 50 to 100%. I am going with the former.

The implementation of the astable model is very similar to the monostable, except that with the duty cycle calculation, there are a few more limitations on the Solver solution. Because of the similarities, we will go through this a little quicker.

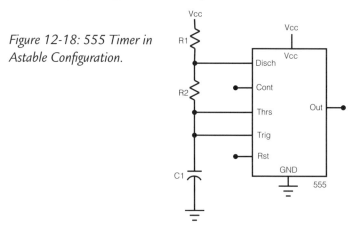

Figure 12-18: 555 Timer in Astable Configuration.

Worksheet Setup

The initial setup of the worksheet appears in Figure 12-19. Cell B5 has been set up for data validation for value from 1 to 999, and cell C5 has been set up for Hz and KHz. Cell E5 has the actual frequency in Hz. Cell B8 has been set up for a data validation of 1–50.

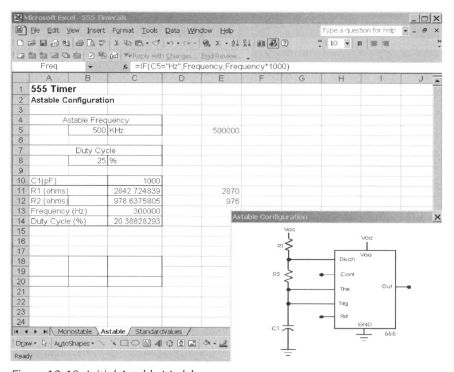

Figure 12-19: Initial Astable Model.

Since the maximum recommended frequency for the 555 is 500 KHz, cells B5 and C5 have been conditionally formatted to turn red for cell E5 greater than 500000. You remember how to conditionally format? No–Select cells B5 and C5, **Format** | **Conditional Formatting** and enter the details as shown in Figure 12-20.

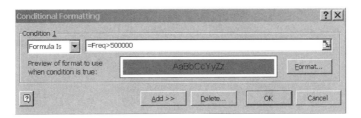

Figure 12-20: Conditional formatting of cells B5 and C5.

Cell C13 contains the formula to calculate the frequency:

$$=1.44/((RES1A+(2*RES2A))*(CAP1A*1e-12))$$

and C14 has the formula for the duty cycle:

$$=(RES2A/(RES1A+(2*RES2A)))*100$$

The macro that is recorded is based on the setup of Figure 12-21. The constraints allow the duty cycle to be ±5% of the desired value and force the resistors to be at least 100 ohms.

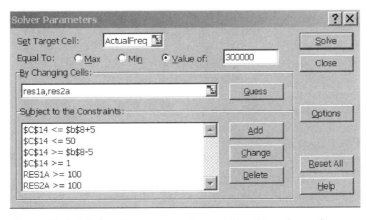

Figure 12-21: Solver Parameters, Assume Non-Negative option set.

When we try to run this macro (before expanding the procedure to vary C1), we notice that the first and third constraints are lost. The macro records them as:

　'*SolverAdd CellRef:="$C$14", Relation:=1, FormulaText:="b8+5"""*

and,

　'*SolverAdd CellRef:="$C$14", Relation:=1, FormulaText:="b8-5"""*

This appears to be yet another minor flaw in Solver and/or its relationship with the Macro Recorder. I would guess that any relative calculation recorded in this way will fail. The solution is simple. Create two cells on the worksheet that perform the calculation. Cell E8 has the formula =b8+5, and cell F8 contains =b8–5. These two cells absolve Solver from doing the calculation and the lines become:

'SolverAdd CellRef:="$C$14", Relation:=1, FormulaText:="$e8"

and,

'SolverAdd CellRef:="$C$14", Relation:=1, FormulaText:="$f8"

This now runs correctly. The listing to date is:

```
Sub SolveAstable()

    SolverReset
    SolverOptions MaxTime:=100, Iterations:=100, Precision:=0.000001, AssumeLinear _
        :=False, StepThru:=False, Estimates:=1, Derivatives:=1, SearchOption:=1, _
        IntTolerance:=5, Scaling:=False, Convergence:=0.0001, AssumeNonNeg:=True
    SolverOk SetCell:="$C$13", MaxMinVal:=3, ValueOf:="300000", ByChange:= _
        "$C$11,$C$12"
    'SolverAdd CellRef:="$C$14", Relation:=1, FormulaText:="$b$8+5"""
    'this line did not record correctly replaced by:
    SolverAdd CellRef:="$C$14", Relation:=1, FormulaText:="$e$8"
    'SolverAdd CellRef:="$C$14", Relation:=3, FormulaText:="$b$8-5"""
    'this line did not record correctly replaced by:
    SolverAdd CellRef:="$C$14", Relation:=3, FormulaText:="$f$8"

    SolverAdd CellRef:="$C$14", Relation:=1, FormulaText:="50"
    SolverAdd CellRef:="$C$14", Relation:=3, FormulaText:="1"
    SolverAdd CellRef:="$C$11", Relation:=3, FormulaText:="100"
    SolverAdd CellRef:="$C$12", Relation:=3, FormulaText:="100"
    SolverOk SetCell:="$C$13", MaxMinVal:=3, ValueOf:=Range("$e$5").Value, ByChange:= _
        "$C$11,$C$12"
    SolverSolve
End Sub
```

Following the same techniques as before, we add the capacitor search. From some experimentation, it also becomes obvious that we need to add an upper constraint to R1 and R2 limiting them to 1 MΩ, and so the completed procedure becomes:

```
Sub SolveAstable()

    Dim varSolveReturn As Variant
    Dim intI As Integer

    Application.ScreenUpdating = False
    For intI = 16 To 63
```

```
'starting at 100pF and limiting at 10uF
    Range("c10").Value = Application.WorksheetFunction.Index(Worksheets("Standardvalues").
Range("D5:D82"), intl)

    SolverReset
    SolverOptions MaxTime:=100, Iterations:=100, Precision:=0.000001, AssumeLinear _
        :=False, StepThru:=False, Estimates:=1, Derivatives:=1, SearchOption:=1, _
        IntTolerance:=5, Scaling:=False, Convergence:=0.0001, AssumeNonNeg:=True
    SolverOk SetCell:="$C$13", MaxMinVal:=3, ValueOf:="300000", ByChange:= _
        "$C$11,$C$12"
    'SolverAdd CellRef:="$C$14", Relation:=1, FormulaText:="$b$8+5"""
    'this line did not record correctly replaced by:
    SolverAdd CellRef:="$C$14", Relation:=1, FormulaText:="$e$8"
    'SolverAdd CellRef:="$C$14", Relation:=3, FormulaText:="$b$8-5"""
    'this line did not record correctly replaced by:
    SolverAdd CellRef:="$C$14", Relation:=3, FormulaText:="$f$8"

    SolverAdd CellRef:="$C$11", Relation:=1, FormulaText:="1000000"
    SolverAdd CellRef:="$C$12", Relation:=1, FormulaText:="1000000"
    'these lines added to limit maximum resistor value to 1M

    SolverAdd CellRef:="$C$14", Relation:=1, FormulaText:="50"
    SolverAdd CellRef:="$C$14", Relation:=3, FormulaText:="1"
    SolverAdd CellRef:="$C$11", Relation:=3, FormulaText:="100"
    SolverAdd CellRef:="$C$12", Relation:=3, FormulaText:="100"
    SolverOk SetCell:="$C$13", MaxMinVal:=3, ValueOf:=Range("$e$5").Value, ByChange:= _
        "$C$11,$C$12"
    varSolveReturn = SolverSolve(True)
    If varSolveReturn = 0 Then
        Exit For
        'force a break
    End If
Next intl

If varSolveReturn <> 0 Then
    SolverSolve
    'rerun with failed values and no update of C1.
End If

    Application.ScreenUpdating = True
End Sub
```

A command button is placed on the worksheet and the associated click will call this procedure. Also, we want to disable the button if the inputs are invalid, so we add the following in the Astable worksheet change event:

```
Private Sub Worksheet_Change(ByVal Target As Range)
    If Range("e5").Value <> 0 And Range("b8").Value <> 0 Then
        Sheet2.cmdAstableSolve.Enabled = True
    Else
        Sheet2.cmdAstableSolve.Enabled = False
    End If
End Sub
```

The result of the calculation is summarized in cells A18 to C23, where the data is formatted into traditional engineering format. For instance, cell A23 contains the formula

=IF(ActualFreq<1000,ROUND(ActualFreq,0)&"Hz",ROUND((ActualFreq/1000),2)&"KHz")

Figure 12-22 shows the completed worksheet.

So there you have it. Quite a lot to say for such a simple circuit!

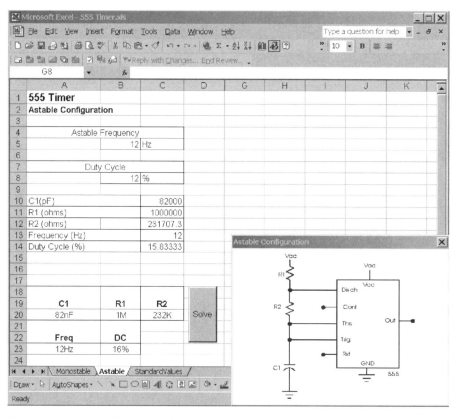

Figure 12-22: Completed worksheet. (Columns E,F are hidden.)

EXAMPLE 13

Purchase Order Generator

Model Description

Assume you work for a small organization and you need to create a purchase order form. These forms will be filled in by hand, but it would be more professional if the forms were numbered in print, rather than scrawled in. This model will allow you to print many copies of the purchase order, incrementing the purchase order number on each sheet by one.

This may seem trivial after some of the models we have produced, but this will allow me to show you how to create an application that hides the fact (well almost!) that it is an Excel application. I know using Excel is nothing to be ashamed of, but there are times when the user does not have sufficient knowledge to open the application and run a macro. All the user will have to do is double-click the icon on the desktop, add one or two numbers on a form and the process is over, bar the printing.

Create a Purchase Order

The first thing to do is to create a Purchase Order (PO) using all the formatting tools at your disposal. Remember, it is possible to insert graphics (**Insert** | **Picture** | select object) if you want to use the corporate logo.

All that you have to do for this model is allocate one cell where the PO number will reside. You should seed this cell with a number, and this number will become the first number in the PO numbering sequence. Figure 13-1 was my attempt.

Once you are happy with the visual effect, you need to see how the worksheet will print out. First, you block the entire area that you want to be printed, and then click on **Files** | **Print Area** | **Set Print Area**. Although you won't be using this directly in this model, it is useful to note that Excel automatically names the block to "Print_Area" for ease of reference. Before we actually print it, you need to size and orient the printout. Click on **File** | **Page Setup**. The dialog box of Figure 13-2 will pop up.

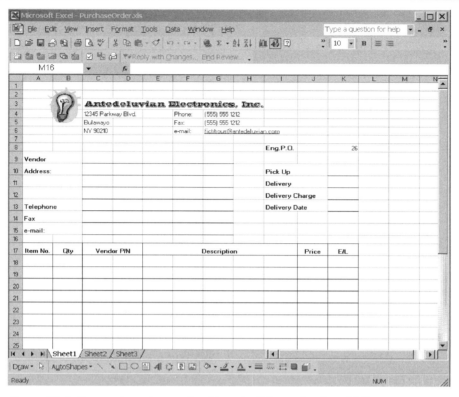

Figure 13-1: Prepared purchase order. Note the PO number in cell K8.

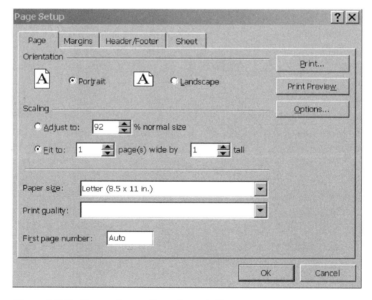

Figure 13-2: Page setup options for printing.

While you are here, note the tab options. You can set the page margins, add a header and footer on every sheet and if you look under the **Sheet** tab (Figure 13-3), you will find that it is possible to create title rows and columns that will repeat on every page when you have a table that spans several pages both horizontally and vertically.

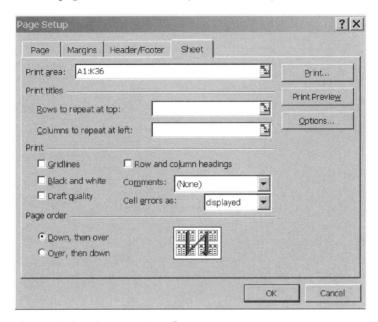

Figure 13-3: Printout options for every page.

If you return to the **Page** tab (Figure 13-2), you will notice a very convenient option which automatically sizes the selected area to print on one (or more) pages. This will be great for the PO model. Select **Print to 1 page** and click on **OK**. If you have more than one printer, also select the printer that should be used.

After printing it out, make any modifications you like to achieve the appearance you are looking for. Remember to save frequently, especially now since the procedures that you are going to be playing with are going to terminate and start Excel automatically.

Print Macro

We need to record a macro that encompasses these printout actions. It is probably a good idea to include the print area definition in the macro to ensure the correct area is printed every time. If you already know how to record a macro, please forgive the repetition. Click on **Tools | Macro | Record new macro** and select a name. Then repeat the print setup process above and then remember to stop recording the macro. The result is:

```
Sub PrintPO()

    Range("A1:K36").Select
    ActiveSheet.PageSetup.PrintArea = "$A$1:$K$36"
    With ActiveSheet.PageSetup
        .PrintTitleRows = ""
        .PrintTitleColumns = ""
    End With
    ActiveSheet.PageSetup.PrintArea = "$A$1:$K$36"
    With ActiveSheet.PageSetup
        .LeftHeader = ""
        .CenterHeader = ""
        .RightHeader = ""
        .LeftFooter = ""
        .CenterFooter = ""
        .RightFooter = ""
        .LeftMargin = Application.InchesToPoints(0.748031496062992)
        .RightMargin = Application.InchesToPoints(0.748031496062992)
        .TopMargin = Application.InchesToPoints(0.984251968503937)
        .BottomMargin = Application.InchesToPoints(0.984251968503937)
        .HeaderMargin = Application.InchesToPoints(0.511811023622047)
        .FooterMargin = Application.InchesToPoints(0.511811023622047)
        .PrintHeadings = False
        .PrintGridlines = False
        .PrintComments = xlPrintNoComments
        .CenterHorizontally = False
        .CenterVertically = False
        .Orientation = xlPortrait
        .Draft = False
        .PaperSize = xlPaperLetter
        .FirstPageNumber = xlAutomatic
        .Order = xlDownThenOver
        .BlackAndWhite = False
        .Zoom = False
        .FitToPagesWide = 1
        .FitToPagesTall = 1
        .PrintErrors = xlPrintErrorsDisplayed
    End With
    ActiveWindow.SelectedSheets.PrintOut Copies:=1, Collate:=True
End Sub
```

There are, no doubt, many superfluous entries in this macro. Unless you are really concerned with the speed of execution or out of a deeper interest in how to code Excel, there is not much point in analyzing it in detail.

User Form

In order to activate the printout of the PO, you will need some operator input. For reasons that I will discuss shortly, you cannot resort to the InputBox technique and you will have to create your own form. In the VBA editor, click on **Insert | User Form**. Then using the tool-box, add two Labels, two Text boxes and two Command buttons, as shown in Figure 13-4.

Figure 13-4: User form creation.

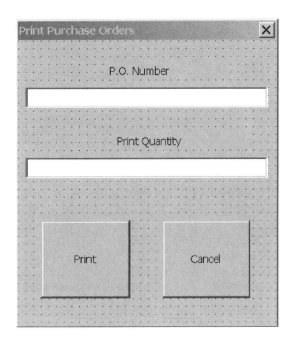

Click on the user form and change the properties in the Properties window. If the properties window is not active, select the object and right click on it and select **Properties**.

Figure 13-5: User form properties.

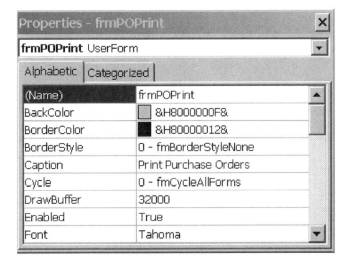

Only three properties need to be changed. The first two are the form name and caption. The third property that I changed is most important to this application¯change the **ShowModal** property to **False**. If a user form is modal, once activated, the user must close the window before proceeding to any other window. An extract from the procedure to run the PO application is:

```
Sub PO()
    'run the application minimized
    Application.WindowState = xlMinimized
    'and now show the form
    frmPOPrint.Show
End Sub
```

Despite the coding order of this procedure, it will show the user form and then proceed to minimize the Excel window (thereby hiding its operation). If the user form is modal, it will not get to the process to minimize the workbook until the user form is terminated. The Excel form will remain visible, something we are trying to avoid. This is the same reason that the InputBox cannot be used since it too, is modal. If the user form is modeless, the form dialog is opened, but the procedure then continues on to minimize the application without waiting for user input. The full code will be discussed later.

You also need to modify the properties of the other controls. You can do this by right-clicking on each control and selecting **Properties.** Actually, if the Properties window is already open, just clicking on the control should do it. As another alternative, you can find the properties by clicking on the Combo box at the top of the Properties window and selecting the object as in Figure 13-6.

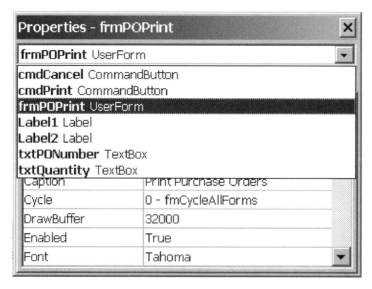

Figure 13-6: Selecting the properties of an object.

All you really need to do is to set up the captions on the object properties, but of course, it does make sense to give them meaningful names. The only other properties of interest are the **TabStop** and **TabIndex**. This affects where the focus (the object where a keyboard action will take place) moves when the **Tab** key is pressed. As each control is placed on the form, it is assigned a TabIndex number. Pressing the **Tab** moves from control to control in the sequence defined by the TabIndex. Changing the numerical order by modifying this property will resequence this order. In some controls, like a label or a form, the **TabStop** is set to **False** automatically. If you don't want the tab to move to a particular control, you can also manually (or in code) set the **TabStop** property to **False**.

When editing a workbook and you want to find the user forms, you will find them listed under the Forms folder in the VBA Explorer. Double-clicking on the form name will show the form.

Initial Procedure

The initial procedure to start the whole process is as follows:

```
Sub PO()
    'run the application minimized
    Application.WindowState = xlMinimized
    'and now show the form

    frmPOPrint.txtPONumber.Text = Range("K8").Value
    'fetch the last value that was printed +1

    frmPOPrint.txtQuantity.Text = "1"

    frmPOPrint.Show
    frmPOPrint.txtPONumber.SetFocus
    'place cursor at the end of the line
    'although not sticly necessary in this case
    'where tab index and tab stop settings ensure
    'this is the first control to gte the focus.
End Sub
```

The first instruction minimizes the Excel window. As discussed earlier, this does not happen immediately, but VBA moves on to the next two instructions which prepare the initial values to be shown in the text boxes of the user form. The initial number for the purchase order is retrieved from the cell K8. The default number of POs is 1. The next value to be written in the next batch of purchase orders that will be printed in the future is stored in cell K8 at the end of the print process (see later).

The fourth instruction displays the user form in all its glory and is followed by an instruction to set the focus to the txtPONumber text box. As it happens in this case, the TextIndex property (combined with the TextStop) is the lowest value so it will be the first object selected. As a result we could have omitted it.

As the application runs, you will see the blank Excel workbook open and then minimize to the task bar. This is the only indication that Excel is running, and requires no user interaction.

Event Actions

Entry of data into the text boxes is handled automatically. There are only four events that you need to consider. They are: click on the **Print** button, click on the **Cancel** button, click on the form terminate button (the "**X**" in the right-hand corner) and workbook startup.

In order to find the location code for the button events, if the user form is visible, right-click on the button and select **View Code** from the menu. The associated click code will pop up immediately. Alternatively, you could right-click on the form name in the VBA Explorer and select **View Code** from that menu. Depending on the view you have chosen for the code window, all events may not be visible by scrolling up and down. Once you have selected the form object, you can locate the command object within it from the Combo box on the left-hand side, and the event from the right-hand Combo box above the code window as shown in Figure 13-7.

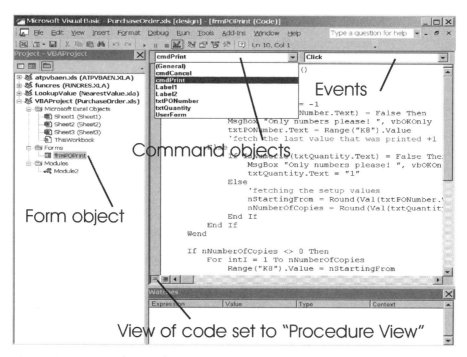

Figure 13-7: VBA code event location.

The **Cancel** click event simply closes the user window and then shuts down Excel. Debugging this can be painful since Excel may have already terminated while you are trying to isolate the problem. You may want to exclude the Excel termination until the application is almost complete.

The code is:

```
Private Sub cmdCancel_Click()
    Me.Hide
    'Unload Application
    Application.Quit
End Sub
```

Find the UserForm_Terminate event and add the exact same code.

Now to consider how to run the application once the **Print** button is clicked:

```
Private Sub cmdPrint_Click()
    Dim intI As Integer

    nNumberOfCopies = -1
    'used to flag accepted data entry

    While nNumberOfCopies = -1
        If IsNumeric(txtPONumber.Text) = False Then
            MsgBox "Only numbers please! ", vbOKOnly
            txtPONumber.Text = Range("K8").Value
            'fetch the last value that was printed +1
        Else
            If IsNumeric(txtQuantity.Text) = False Then
                MsgBox "Only numbers please! ", vbOKOnly
                txtQuantity.Text = "1"
            Else
                'fetching the setup values
                nStartingFrom = Round(Val(txtPONumber.Text), 0)
                nNumberOfCopies = Round(Val(txtQuantity.Text), 0)
            End If
        End If
    Wend

    If nNumberOfCopies <> 0 Then
        For intI = 1 To nNumberOfCopies
            Range("K8").Value = nStartingFrom
            Call PrintPO
            nStartingFrom = nStartingFrom + 1
        Next intI
        'save the values in the workbook
        'to prevent the "ARe you sure question
        'when the application closes
        Range("K8").Value = Range("K8").Value + 1
        'saving the next number to be printed
        ActiveWorkbook.Save
```

```
    End If

    Me.Hide
    'Unload Application
    Application.Quit
  End Sub
```

Two public variables are used and are declared in the general declarations area of the modules and are:

```
  Public nStartingFrom As Integer
  Public nNumberOfCopies As Integer
```

The procedure first analyzes the input. If the user has modified the PO number in the P.O.Number box, this becomes the start of a new numbering sequence. The numbering sequence that is stored on K8 is erased, and the next number to be printed in the new numbering sequence is stored back to cell K8 for the next batch of purchase orders that will be printed. If there are illegal inputs (non-numeric characters in this case), a message box is used to alert the user of the error. Since Excel is already minimized, the modality of the message box does not interfere with the operation. If the input is legitimate, the print process is repeated for the number of times entered in the txtQuantity text box. In this process, the PO number is stored at cell K8 and then the form is handed over to the Windows print handler. The PO number is incremented and the process is repeated until the required number of sheets has been printed.

Once the print process is completed, the value of the next PO number is placed in the worksheet and the worksheet is saved. The reason for this is twofold. First, the worksheet is the repository of the PO number to be used next time. Second, as you know from regular operation, exiting a workbook if it has been modified will result in the user prompt with the option of saving the worksheet. Saving it first disposes of this message allowing the application to close without any user response needed.

Finally the application unloads the form and quits the application.

Auto Startup

To run this procedure as the worksheet is open we place the following in the worksheet open event:

```
  Private Sub Workbook_Open()
    Call PO
  End Sub
```

Incidentally, if you need to debug this application, open Excel and then hold the **Shift** key down while opening the application file (PurchaseOrder.xls, in this case). Also, remember that **<Ctrl> + <Break>** stops VBA code from executing.

Running PurchaseOrder

All that is left to do is close Excel, and place a shortcut to the workbook "PurchaseOrder.xls" on the desktop. Double-clicking on the icon (and of course you can change the default icon) will run the whole application with no user knowledge of Excel and with little awareness that Excel is in fact used for the application.

If the print function should fail, Windows will take care of that, alerting the user through the standard error messages outside of this application. Figure 13-8 gives an idea of the end result.

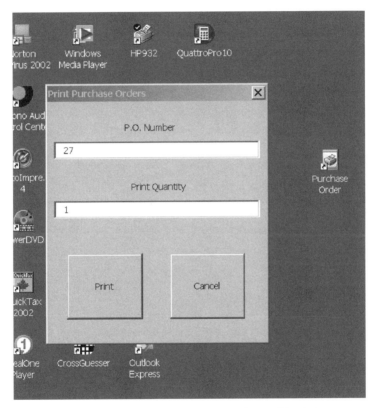

Figure 13-8: Running the application. Note the workbook icon on the right. No sign of Excel anywhere!

EXAMPLE 14

Interface to a Digital Multimeter Using a Serial Port

Model Description

Many electronic instruments are supplied with a computer interface. Several manufacturers, like Agilent, even provide utilities that allow the instrument to operate within Excel. So why do I want to reinvent the wheel? Well, my experience of these interfaces is that they are limiting. First, unless you use a GPIB interface, you are stuck using one instrument at a time and, as you saw in Example 1, a simple automated test would need at least two instruments, one to stimulate the input and one to measure the output. Second, mixing Excel drivers for instruments from different manufacturers within Excel may generate conflicts between them. Third, the instrument you want may not have a driver for Excel, and fourth, being an engineer there is a high probability that you will end up designing your own piece of equipment with some kind of interface port on it.

This model will create an RS-232 interface through Excel with the Radio Shack 220-0812 Digital Multimeter (DMM). Although the RS-232 port on PCs is slowly being edged out, many instruments still retain this communications capability and probably will for some time in the future. There are USB-to-serial port adapters on the market, so the communications ability will still exist. Using multiple serial ports, or by using a serial port expander (also controlled through the serial port) like the 232SS2 from B&B Electronics will allow the use of multiple instruments. Of course you can convert to RS-485 and create your own multidrop network.

The Radio Shack DMM is probably not the most accurate meter on the market, but it does have a serial interface, the price was right for my experimentation and most of the protocol is published. In addition, the interface is unusual enough to justify the creation of our own software, yet simple enough not to blur the explanation that I will try to give. The DMM does come with its own software to chart and record any reading, but it runs outside of Excel. It is not my intention to duplicate this functionality; I merely want to use a particular DMM setting to take a reading periodically, introduce that into a worksheet and process it.

For a description of different methods of interfacing to the serial and parallel ports, see Appendix B. I have chosen to use the Visual Basic MSCOMM driver because it will allow us to write almost any serial interface, provided we can get a definition of the protocol that a particular instrument uses.

DMM Interface Protocol

The Radio Shack DMM communicates at 4800 baud, using 8-bit data, no parity and 1 stop bit. It transmits a burst of 9 bytes about every 200 mS with no handshaking at all. Some of this was deduced with the aid of an oscilloscope. It is not stated anywhere, but it appears as if the RS-232 interface is powered from the DTR (must be at +12V) and RTS (must be at –12V) for the serial port of the DMM to function. Aside from the first and last bytes, each bit in the packet message represents an LCD element on the display. For instance, a seven segment digit 2 (see Figure 14-1 and Table 14-1) would result in a byte 0xb5 for that digit. Where a digit contains the decimal point indicator, the dot actually precedes the digit. Even the "m" and "V" in "mV" are represented by two bits.

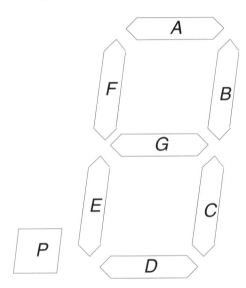

Figure 14-1: 7 Segment identification.

Packet #	Description	LCD	Bit7	Bit6	Bit5	Bit4	Bit3	Bit2	Bit1	Bit0
0	Mode	----	Mode (see Table 14-2)							
1	Units1	LCDF-0x3F	Hz	Ω	K	M	F	A	V	m
2	Units2	LCDE-0x3E	u	n	dBm	S	%	hFE	REL	MIN
3	Least Sign. Digit	LCDD-0x3D	D	C	G	B	P	E	F	A
4	2nd digit	LCDC-0x3C	D	C	G	B	P	E	F	A
5	3rd digit	LCDB-0x3B	D	C	G	B	P	E	F	A
6	Most Sign. Digit	LCDA-0x3A	D	C	G	B	P	E	F	A
7	Extra	LCD9-0x39	BEEP	DIODE	BAT	HOLD	-	~	RS232	AUTO
8	Check	----	Checksum + 0x57							

Table 14-1. Bit identification in transmission packet.

Mode	Function
0	DC V
1	AC V
2	DC uA
3	DC mA
4	DC A
5	AC uA
6	AC mA
7	AC A
8	OHM
9	CAP
10	HZ
11	NET HZ
12	AMP HZ
13	DUTY
14	NET DUTY
15	AMP DUTY
16	WIDTH
17	NET WIDTH
18	AMP WIDTH
19	DIODE
20	CONT
21	HFE
22	LOGIC
23	DBM
24	EF
25	TEMP

Table 14-2: Mode decoder.

Pin	Name	Direction
1	DCD- Data Carrier Detect	Into PC
2	RD- Received Data	Into PC *
3	TD- Transmitted Data	Out of PC
4	DTR- Data Terminal Ready	Out of PC*
5	SG- Signal Ground	n/a*
6	DSR- Data Set Ready	Into PC
7	RTS- Request To Send	Out of PC*
8	CTS- Clear To Send	Into PC
9	RI- Ring Indicator	Into PC

Table 14-3: Pinout for 9-way D-Sub male connector on a PC.
* Only these signals are used in this example.

Unfortunately, Radio Shack provides little in the way of details on how the checksum is calculated. I have tried a number of different combinations including using the cryptic information in the "LCD" column of Table 14-1, but I could not get to the same result that they do.

Having looked at a few DMMs before starting this example, I also get the feeling that obtaining protocol information from some other instrumentation manufacturers may be difficult.

MSComm32

In order to work with this example, you will need to have the ActiveX control MSComm32 installed and registered. This may not be an easy task, so see Appendix B for details. If you intend to create applications with the serial port, then I am sure you will benefit from two other books: *Serial Port Complete* by Jan Axelson and *Visual Basic Programmer's Guide to Serial Communications* by Richard Grier. (See the references at the end of the book.)

Before you can access MSComm32 (provided it is installed and registered in Excel), it must be placed as an object in a worksheet. This can be done in two ways. It can be placed in the worksheet directly, or on a form in VBA. The only difference is the actual location when you actually use the function. In this model, I have placed the control on UserForm1 (we will get to it later). The reference to the control is entered as:

UserForm1.MSComm1

If placed in a worksheet, the UserForm1 will be replaced by the sheet name.

To place the MSComm32 control on a worksheet, first make sure that the Control Toolbox is visible in the worksheet. If not, click on **View** | **Toolbars** | **Control Toolbox** . Then click on the **More controls** button and select **Microsoft Communications Control** as in Figure 14-2.

Then click and drag a rectangle on the worksheet and a control with a telephone icon should appear as in Figure 14-3. Note that when the Design Mode button (the icon is a set square) on the Control Toolbox is in the Run Mode, the MSComm control is invisible.

Figure 14-2: Placing more controls not on toolbar.

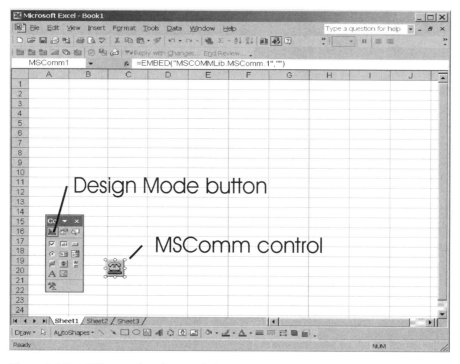

Figure 14-3: MSComm placed on worksheet.

An alternate way (and the way I have done it in this model), is to place the control on a user form. In VBA (**<Alt> + <F11>**, or **Tools | Macro | Visual Basic Editor**) click on **Insert | Userform**. If the toolbox is not visible, make sure that the userform is selected by clicking on it. If it still is not visible, enable it through the **View | Toolbox** sequence. Right-click on the toolbox (if the MSComm telephone icon is not already there) and select **Additional Controls** from the pop-up menu. Then select **Microsoft Communications Control** as in Figure 14-4. Click on **OK** and the telephone icon will appear on the toolbox.

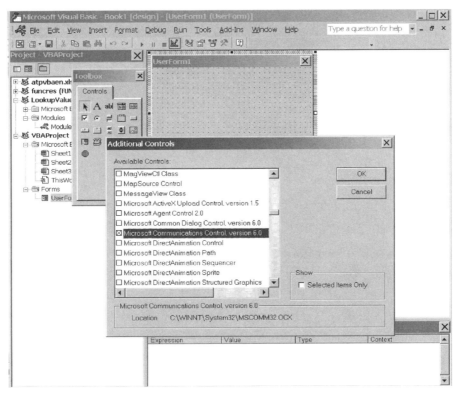

Figure 14-4: Adding MSComm Control to the tools.

In Parenthesis: *Additional Controls*

When you have some spare time, try enabling some of the controls that appear in the Additional Controls list. I don't know which are standard with Office. Some may get added if Visual Basic is installed, as well as from many other applications. On my system I managed to find a Level Slider Control, Microsoft Progress Bar Control, and a Knob Control.

Place the MSComm control by clicking on the MSComm icon on the toolbox and then click and drag a rectangle on the user form and the icon will appear as in Figure 14-5.

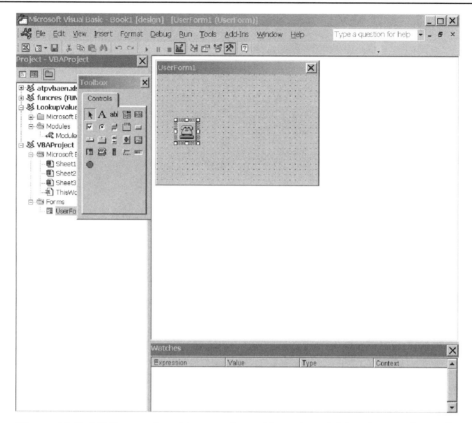

Figure 14-5: MSComm placed in user form. Note the additional controls on the toolbox.

In Parenthesis: *MSComm Properties*

Most Universal Asynchronous Receiver/Transmitters (UART) have a number of setup registers. The device used on the PC has its fair share. The only way to set these up in MSComm is by accessing the properties of the control. Here is a description of most of those properties.

CommPort: *object.**Commport**[=value]*

There can be up to sixteen COM ports on a PC, denoted COM1 to COM16. This property must be set prior to opening a port and must be set to a port that does exist in the hardware map or an error 68 will occur.

PortOpen: *object.**PortOpen**[=value]*

Before a port is used, it must be opened (or an error 68 will occur) by setting this property to TRUE. Its status can be interrogated by checking this property. Obviously setting it to FALSE closes the port. The port should be closed as a matter of good design practice when the application is shut down.

Settings: *object.**Settings** [= value]*

This property sets the standard communications properties: baud rate (BBBB), parity (P), number of data bits (D), and number of stop bits (S). These are fed as a string in the format:

"BBBB,P,D,S"

BBBB, the baud rate, can have one of the following values 110, 300, 600, 1200, 2400, 4800, 9600, 14400, 19200, 28800, 38400, 56000, 128000, 256000. This does not guarantee that the hardware can work at the higher baud rates.

P, the parity setting can be E for Even, M for Mark, N for None, O for Odd, and S for Space. Using the M and S, it is possible to implement the 9-bit Intel style UART (see reference 65).

D, the data bits setting, can have a value of 4,5,6,7 or 8.

S, the stop bit settings, can be 1, 1.5, or 2

Handshaking: *object.**Handshaking**[=value]*

There are two types of handshaking protocol between RS-232 interfaces. They are needed with relatively high speed data rates to prevent the loss of data. XON/XOFF embeds codes in the serial data stream to start and stop transmission. An alternative uses the RTS/CTS signal lines. There are four possible values: comNone (no handshaking used), comXOnXOff, comRTS, and comRTSXOnXOff (both XOn/XOff and RTS/CTS).

RThreshold: *object.**RThreshold** [=value]*

The MSComm control can work through interrupt actions (see OnComm decription). This property will generate an interrupt after the set number of characters have been received. Fixing this value to 0 disables the onComm interrupt.

InBufferCount: *object.**InBufferCount***

As data is received, it is accumulated for the number of characters in RThreshold and then the OnComm event occurs. The number of characters stored is available by reading this property.

InputLen: *object.**InputLen**[=value]*

As data is received, it is accumulated for the number of characters in RThreshold until the OnComm event. When reading the data using the Input property (see later in this box), InputLen determines how many characters are read. This is good for fixed length messages. A value of 0 will read out all the received bytes. If an attempt is made to access the received data before this number of bytes have been received a zero length string is returned ("").

DTREnable: object.***DTREnable**[=value]*

Setting this to FALSE disables the DTR line (–12V). Setting this to TRUE enables the DTR line (+12V). When the port is closed, the output is changed back to –12V.

RTSEnable: object.***RTSEnable****[=value]*

Setting this to FALSE disables the RTS line (–12V). Setting this to TRUE enables the RTS line (+12V). When the port is closed, the output is changed back to –12V.

CommEvent: *object.****CommEvent***

The OnComm event is a blanket interrupt for any event within the MSComm control. This property defines which event actually happened. It is described more fully in the "In Parenthesis: OnComm" box in this example.

InputMode: *object.****InputMode****[=value]*

Data can be received as text or binary format. This property can be set to comInputMo-deText or comInputModeBinary.

Input: *object.****Input***

Reading this property removes a stream of data from the input buffer. If the data is in text format (from InputMode property), it is returned in a variant. If it is binary, it is returned as an array of bytes in a variant. Got that? Never mind, just follow the example, or read *Serial Port Complete*, Chapter 4.

Output: *object.****Output****[=value]*

This is how data is transmitted. If the data to be transmitted is text (only ANSI strings), then the output is a variant that contains a string. If the data is binary, then we have to use a variant that contains a byte array. Confused? Me too. Example 14 (this example) only deals with data reception. Example 16 will have both transmission and reception. Of course, the aforementioned books also covers this.

SThreshold: *object.****SThreshold**** [=value]*

The MSComm control can work through interrupt actions (see OnComm decription). This property will generate an interrupt while there are less than SThreshold characters in the buffer.

Initializing the Serial Port

Because I don't know how to calculate the checksum in this protocol, I have had to resort to the fact that there is about 200 mS between transmissions in order to synchronize the readings. Initially, we will set the serial port to look for single characters and every time it sees one, a timer is reset. When the timer exceeds 100 mS, the serial port is initialized to read packets of 9 characters. Let's create the code to do this.

In VBA, click on **Insert | Module**. In the General Declarations, add the following global variables. (I know this will incur the wrath of software purists. Please forgive me for showing you my bad habits derived from programming microcomputers when the world was young and 64 bytes was all the RAM I had.)

```
Public bSeekingSync As Boolean
'used to indicate that the program is
'looking to synchronise the bytes
Public nFrozenTime As Single
'used to timeout to allow sync
Public iDMMMessage(9) As Byte
Public bEndFlag As Boolean
'used to denote stop button

'input buffer
Public bBlocReceived As Boolean
'indicates a 9 byte block has been recived
Public iInputPointer As Integer
```

Add a procedure called PollDMM as follows:

```
Sub PollDMM()
    Dim i As Integer
    bEndFlag = False
    'set up port to read
    'one byte
    With UserForm1.MSComm1
    'this is just shorthand to save writing
    'UserForm1.MSComm1.xxxx= nnn
    'every time.
        .CommPort = 1
        .PortOpen = True
        .Handshaking = comNone
        'no handshaking required
        .Settings = "4800,N,8,1"
        'set for 4800 baud, no parity, 8 bits, 1 stop bit
        .InputMode = comInputModeBinary
        'binary data returned in variant array
        .InputLen = 1
        'set to one character when read by Input
        .RThreshold = 1
        'number of characters to receive before
        'generating on comm event
        .DTREnable = True
        .RTSEnable = False
    End With
    bSeekingSync = True
    'indicate that we are looking for sync
    nFrozenTime = Timer
    'initialize timer
    bBlocReceived = False
    'and that there is no valid block
```

```
        While (Timer - nFrozenTime) < 0.2
           'waiting untill there is a dead time
           DoEvents
        Wend

        'now to format port to read 9 bytes
        bSeekingSync = False
        UserForm1.MSComm1.RThreshold = 9
        'interrupt after 9 bytes
        UserForm1.MSComm1.InputLen = 0
        'transfer all received bytes when read

        While bEndFlag = False
           DoEvents

           If bBlocReceived = True Then
              'block has been received.
              'write them to sheet1

              Sheets("Sheet1").Select
              For i = 0 To 8
                 Cells(i + 1, 1) = iDMMMessage(i)

                 Cells(11, 1) = iCounter
                 Cells(12, 2) = TestP
              Next i
              bBlocReceived = False
              'clear block and start looking again
           End If
        Wend
     End Sub
```

Initially, the serial port is opened and configured for single-byte reception using the "with" construction. At the end of the port setup, some values are initialized. *bSeekingSync* is a flag that indicates to the OnComm interrupt that we are looking for a sync signal. *nFrozen* is set to the current time in seconds. *bBlocReceived* is cleared, but is not used in the synchronization process. The intention is that during the interrupt caused by a character reception, if *bSeekingSync* is TRUE, then the latest time is stored on *nFrozen*, thereby resetting the counter. In the background loop in PollDMM, the program waits for the time to expire at 200 mS. Pay attention to the DoEvents instruction—without it there will be no OnComm interrupt.

In Parenthesis: *Timer*

The Timer function in VBA returns the number of seconds since midnight. It has a resolution of 10 mS. If this (or any) program is going to run late at night, you need to allow for this rollover.

In Parenthesis: *DoEvents*

Running loops in VBA where there are no screen updates or other system calls, can result in the program consuming a large portion of the system's resources and certain processes may not run. During these loops, it is advisable to execute a DoEvents instruction which allows Windows to handle all the other processes going on.

The OnComm event is an interrupt routine that occurs for almost any status change of the UART. The property CommEvent defines the reason for the interrupt. See "In Parenthesis: OnComm" for a list of the possible values of the property. It is possible to sort out the cause of the interrupt and to service it by using Select Case construct within the OnComm event.

Go to the MSComm code window. You can do this in two ways. In the VBA Explorer, double-click on the UserForm1 folder (or wherever the control is stored). Right-click on the MSComm telephone icon, and select **View Code** from the pop-up window. A quicker way is to right-click on the UserForm1 folder in the VBA Explorer, and select **View Code** from there. This will take us straight to the OnComm event. The code we will use is:

```
Private Sub MSComm1_OnComm()

    Dim Dummy As Variant
    Dim RXbytes() As Byte
    Dim il As Integer

    Select Case MSComm1.CommEvent
        Case comEvReceive
            If bSeekingSync = True Then
            'looking for sync.
                nFrozenTime = Timer
                'refresh time since a character
                'has been seen
                Dummy = MSComm1.Input
                'unload the data
            Else
                'here we must read 9 bytes when ready
                Dummy = MSComm1.Input
                RXbytes() = Dummy
                For il = 0 To 8
                    iDMMMessage(il) = RXbytes(il)
                Next il
                bBlocReceived = True

            End If
        Case Else
    End Select

End Sub
```

Notice in the first part of the OnComm event, while bSeekingSync is TRUE, that timer gets refreshed and the data buffer emptied.

If we look at the PollDMM procedure again, we see that once the synchronization is found, the serial port is reconfigured to handle 9 bytes of data at a time. When the 9 bytes are received, they are actually read in the else clause of the OnComm procedure. A flag is set for the PollDMM procedure to indicate there is valid data available to be processed in the looping section of PollDMM.

The background loop scans two Boolean variables. The first, *bBlocReceived*, has just been discussed. The second, *bEndFlag*, is a flag to indicate that the process should halt. It is generated from a Command button, which we will introduce shortly. Once a block of data is received, it is placed on Sheet1 using the Cells instruction. Just prior to this, the code sets the active sheet to Sheet1 using a *Sheets("Sheet1").Select* statement.

In Parenthesis: *OnComm Event*

It is possible to use the serial port without resorting to the interrupts used by the On-Comm event. Given that the operating system can have a lot to do, and can be fairly unpredictable in doing it, it is safer, in my opinion, to rely on the interrupt structure. The OnComm event can be triggered by the following occurrences. The names are actual VB constants that can be used in the case statements.

Normal Operation:

comEvCD: change in state of the CD input signal.
comEvCTS: change in state of the CTS input signal.
comEvDSR: change in state of the DSR input signal.
comEvRing: change in state of the RI input signal.
comEvReceive: The receive buffer has RThreshold characters stored.
comEvSend: The transmit buffer has less than SThreshold characters stored.
comEvEOF: EOF (End of File) character (0x1A) received.

Exceptions:

comEventBreak: break signal received.
comEventFrame: framing error detected.
comEventOverrun: overrun error detected.
comEventRxOver: more than RThreshold characters received.
comEventRxParity: parity error detected.
comEventTxOver: more than SThreshold characters placed in buffer.
comEventDCB: unexpected error retrieving Device Control Block from port.

Before we attempt to run this, let's place two Command buttons on Sheet1. **View** | **Toolbars Control Toolbox** brings up the correct toolbox. Place two buttons and change the names to *cmdStart* and *cmdStop*. Then change the captions to Stop and Start. In their click events, let's add the code:

```
Private Sub cmdStart_Click()
    cmdStart.Enabled = False
    cmdStop.Enabled = True
    Call PollDMM

End Sub

Private Sub cmdStop_Click()
    Call Module1.ForceStop

End Sub
```

In module1 we add the code to stop the process:

```
Sub ForceStop()
    UserForm1.MSComm1.RThreshold = 0
    'disable interrupts- sometimes the click happens
    'while the inrerrupt is being serviced
    'if this isn't here the click may have no effect
    'in those circustances
    UserForm1.MSComm1.PortOpen = False
    'close the port
    Sheet1.cmdStart.Enabled = True
    Sheet1.cmdStop.Enabled = False
    bEndFlag = True

    End Sub
```

Since we disable the buttons alternately, there is no need at the moment to check if the serial port is open or closed (since a repeat attempt to do either will trigger a fault). We also need to add an initialize procedure in the workbook open event to allow the process to start.

```
Private Sub Workbook_Open()
Sheet1.cmdStart.Enabled = True
Sheet1.cmdStop.Enabled = False
End Sub
```

When working the RS-232, we are always faced with the question as to whether the two devices are DCE or DTE. What it boils down to is whether to use a null modem cable or a straight through cable. We use the latter (supplied with the DMM) in this case. First, connect the DMM to the PC Com port with the cable supplied with the meter. Now turn the DMM on to VDC and set the communications mode to on by simultaneously pressing the Select and Range buttons.

In Excel, click on the Start button and watch the numbers change. Change the function selector on the DMM and observe the effects. It's all downhill from here, but we still have quite a bit to cover. Figure 14-6 is how the worksheet should appear with the DMM reading as in Figure 14-7.

Try stopping the process by clicking on the Stop button. Typically, this is very difficult since it appears that the OnComm interrupt masks the click. I don't know why, but after some trial and error, I found that menu bars are not subject to the same problems so I modified the program to run from a Menu Bar.

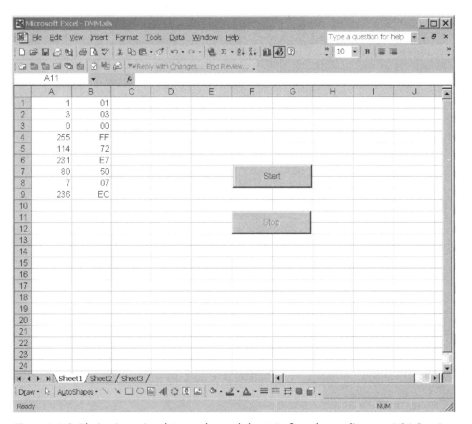

Figure 14-6: Placing incoming data on the worksheet. In fact, the reading was 164.8 mA.

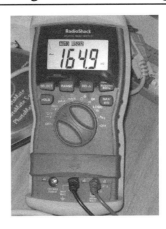

Figure 14-7: Setup of DMM being read into worksheet.

I have left this spreadsheet stored on the CD-ROM as *DMM1.xls*, in case you wish to experience the frustration on having to click many times on the stop button just to get it to be seen by the PC.

Before we add the toolbar, let's get rid of the Command buttons. If the Control Toolbox is not visible, make it so by **View** | **Toolbars** | **Control Toolbox**. Make sure the Design Mode button (the set square) is active. Right-click on each button and select **Cut** from the pop-up menu. Unfortunately, this does not remove the code we have written referring to these ex-buttons. We have to go to VBA and in the VBA Explorer, right-click on **Sheet1** and select **View Code**. Delete the code for both buttons and then in a similar fashion view the code for the workbook and delete the code for the workbook open event.

Now let's create a toolbar. In Sheet1, click on **Tools** | **Customize** and select the **Toolbars** tab. Click on the **New** button and we should see something like Figure 14-8.

After clicking on the **OK** button, this toolbar will be added to the list. (It will be available to any application if it is enabled by placing a check next to the name.) A small toolbar appears with no buttons on it. To add a button, click on the **Commands** tab in the Customize dialog box. Find the entry **Macros** in the Categories window and then click and drag the Custom button to the DMM toolbar. Repeat the process of adding a Custom button and you will have two smiley-faced buttons within the bar. While keeping the Customize dialog box open, right-click on one of the smileys and modify the **Name** to *&Acquire*. The ampersand places a line under the following letter and can be used for a keyboard shortcut. Unfortunately, this shuts the pop-up window and we have to right-click several times to complete the setup. Check next to the **Image and Text** and **Change the Button Image** to the picture of a running man. Assign the macro PollDMM with this button. Change the second button to a Stop button in a similar fashion, assigning the macro ForceStop to it. See Figure 14-9.

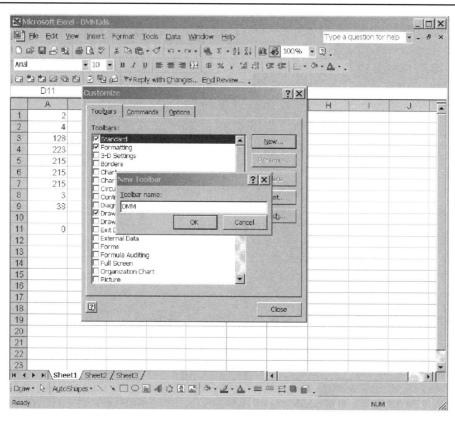

Figure 14-8: Adding a toolbar named DMM.

Figure 14-9: Setting up toolbar button properties.

The toolbar should look like Figure 14-10.

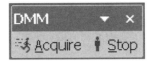

Figure 14-10: Resulting toolbar.

In Parenthesis: *Custom Toolbar Limitation*

The toolbar, once created, will always appear in all workbooks as they are opened. They can be turned off using the "**X**" in the top right-hand corner. Toolbars can be re-enabled from the **Customize** dialog under the **Toolbars** tab (see Figure 14-8). Finding the toolbar and checking the box next to it will re-enable it.

By default, toolbars are stored with Excel and not with the workbook. They can be embedded into the workbook from the **Customize** dialog under the **Toolbars** tab (see Figure 14-8). Click on the **Attach** button, and in the resulting dialog select the desired toolbar that is listed in the Custom Toolbars window and click on the **Copy>>** button to copy the toolbar to the Toolbars in workbook window. Click on the **OK** button followed by the **Close** button and save the workbook.

As a direct upshot of this, the macro associated with a menu button is tied to a particular workbook. If that workbook is not open, it will be opened. This can be positive since it will automatically open a workbook when the control is clicked without any effort on our part. However, in our current example it is problematic. As we go, I am saving different versions of the same workbook and the macros, although they are named the same, are different. Make sure that the macros the buttons refer to are in the current workbook. In order to do this, make sure the Customize dialog is open, right-click on the button and edit the cell to the associated macro.

We also need to edit both macros to make sure that we don't try to open an open port or to close a closed port. The beginning of PollDMM becomes:

```
Sub PollDMM()
    Dim i As Integer
    bEndFlag = False
    'set up port to read
    'one byte
    With UserForm1.MSComm1
    'this is just shorthand to save writing
    'UserForm1.MSComm1.xxxx= nnn
    'every time.
        .CommPort = 1
        '.PortOpen = True
        'enabled in a few lines to allow check if open
```

```
                .Handshaking = comNone
                'no handshaking required
                .Settings = "4800,N,8,1"
                'set for 4800 baud, no parity, 8 bits, 1 stop bit
                .InputMode = comInputModeBinary
                'binary data returned in variant array
                .InputLen = 1
                'set to one character when read by Input
                .RThreshold = 1
                'number of characters to receive before
                'generating on comm event
                .DTREnable = True
                .RTSEnable = False
            End With
            'add check if port is open
            If UserForm1.MSComm1.PortOpen = False Then
                UserForm1.MSComm1.PortOpen = True
            End If
```

Note that the PortOpen line of code within the With construction has been commented out. There is a check afterwards if the port is already open. Similarly, let's modify ForceStop as follows:

```
        Sub ForceStop()
            UserForm1.MSComm1.RThreshold = 0
            'disable interrupts- sometimes the click happens
            'while the interrupt is being serviced
            'if this isn't here the click may have no effect
            'in those circumstances
            If UserForm1.MSComm1.PortOpen = True Then
                UserForm1.MSComm1.PortOpen = False
            End If
            'close the port
            bEndFlag = True
        End Sub
```

At this time, we can also add a call to ForceStop in the workbook deactivate event so that the port is closed when the application is over.

Conversion of DMM Display to Data

I have decided to ignore all the non-numeric outputs that are possible, like the units. I don't see any point in trying to recreate the universal DVM input since an interface that we are trying to create would likely be to gather a single range of data. I will also presume that since the user has to manually invoke the RS-232 interface, he or she can just as easily set the unit to a particular range (anything but autorange). I hope you find this approach reasonable. It will certainly make the code shorter and more understandable.

In Module1 of VBA, enter the following function:

```
Public Function sCreateDigit(i7Segment As Integer) As String
    Dim bDecPnt As Boolean
    Dim sRetVal As String
    Dim iTemp As Integer
    'checking for decimal point
    'see text for description on And
    iTemp = i7Segment And 8
    If iTemp <> 0 Then
        bDecPnt = True
        i7Segment = i7Segment And 247
        'clearing DP

        sRetVal = "."
        'initiating string with dp
    Else
        bDecPnt = False
        sRetVal = ""
    End If

    'using lookup table for possible characters
    Select Case i7Segment
        Case 215
        '0xd7= 0
            sCreateDigit = sRetVal & "0"
        Case 80
        '0x50 =1
            sCreateDigit = sRetVal & "1"
        Case 181
        '0xb5=2
            sCreateDigit = sRetVal & "2"
        Case 241
        '0xf1=3
            sCreateDigit = sRetVal & "3"
        Case 114
        '0x72=4
            sCreateDigit = sRetVal & "4"
        Case 227
        '0xe3=5
            sCreateDigit = sRetVal & "5"
        Case 231
        '0xe7=6
            sCreateDigit = sRetVal & "6"
        Case 81
        '0x51=7
            sCreateDigit = sRetVal & "7"
```

```
        Case 247
        '0xf7=8
            sCreateDigit = sRetVal & "8"
        Case 243
        '0xf3=9
            sCreateDigit = sRetVal & "9"
        Case Else
            sCreateDigit = "F"
    End Select
  End Function
```

This function is applied to each 7-segment digit, returning a text value for the associated number. I am using strings because the concatenation is easy and we can then convert to a numeric value. If the decimal point appears, it precedes the digit in the text string.

Visual Basic unfortunately uses the same word for both a logical and a bitwise AND. It is taken in context, so that you cannot look for a bit in an integer (as in the above code) by entering *if i7segment AND 8* since VB would evaluate i7segment which is either true or false (nonzero or zero), and logically AND it with the number 8 (which will always be true) and the meaning will not be the same as a bitwise AND. The way around this is to ascribe the operation to a variable as in *iTemp=i7segment And 8* and then have a conditional test for iTemp.

Enter the formula:

    =sCreateDigit(A4)

in cell C4. Copy the cell to C5, C6 and C7. In cell C11, enter the formula:

    =C7 & C6 & C5 & C4

which concatenates the digits and creates a number that should agree with the instantaneous display of the DMM.

In cell C12, the formula:

    =value(c11) converts the string to a number. We could format C11 (and I have) to resemble an LED output by changing the size and color of the cell text format.

Analog Meter Chart

Despite the fact that I said I did not want to recreate a DVM input, there is an interesting way of applying a pie chart to make it look like an analog meter. While it is hardly Labview®, it is certainly a lot cheaper.

Before we start, it might be easier if you looked at Figure 14-19 to get an idea of what we are trying to implement. The full 360 degrees of the pie chart are broken into three sectors. The first sector of the pie chart is obviously our reading as a percentage of full scale. The second sector is the full-scale reading minus the actual reading, and the third sector is the balance of the pie chart. The meter will operate through an angle of 150 degrees (which we can make programmable by entering it in a cell), so we need to scale the ratios according to this. Take a look at the formulas in Figure 14-11.

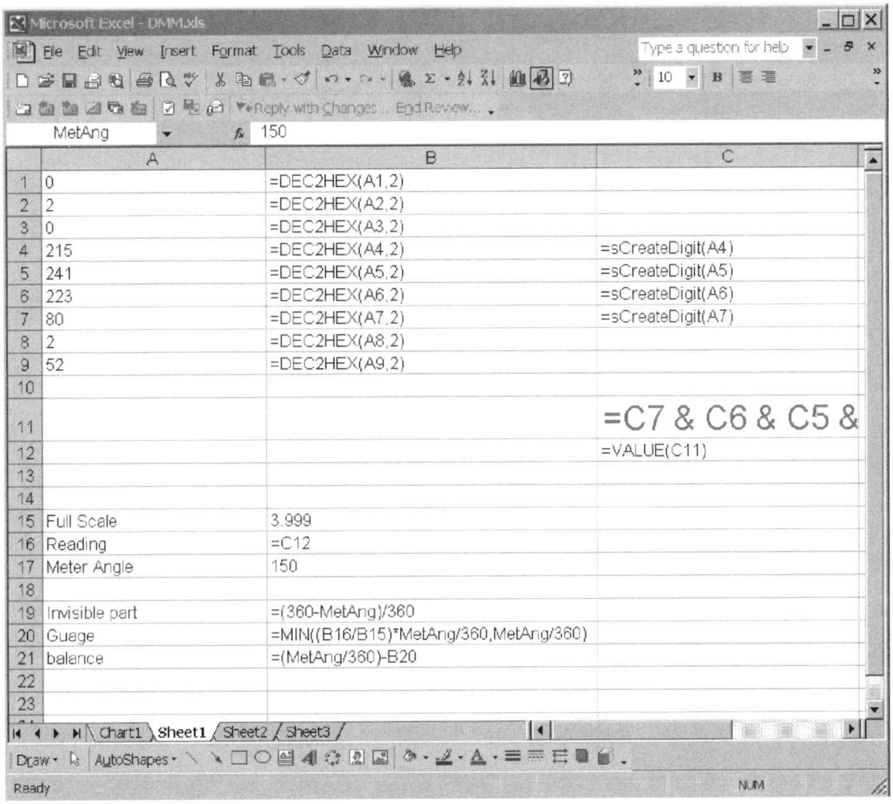

	A	B	C
1	0	=DEC2HEX(A1,2)	
2	2	=DEC2HEX(A2,2)	
3	0	=DEC2HEX(A3,2)	
4	215	=DEC2HEX(A4,2)	=sCreateDigit(A4)
5	241	=DEC2HEX(A5,2)	=sCreateDigit(A5)
6	223	=DEC2HEX(A6,2)	=sCreateDigit(A6)
7	80	=DEC2HEX(A7,2)	=sCreateDigit(A7)
8	2	=DEC2HEX(A8,2)	
9	52	=DEC2HEX(A9,2)	
10			
11			=C7 & C6 & C5 &
12			=VALUE(C11)
13			
14			
15	Full Scale	3.999	
16	Reading	=C12	
17	Meter Angle	150	
18			
19	Invisible part	=(360-MetAng)/360	
20	Guage	=MIN((B16/B15)*MetAng/360,MetAng/360)	
21	balance	=(MetAng/360)-B20	
22			
23			

Figure 14-11: Formulas needed for analog meter. Note cell B17 has been named MetAng.

Obviously the sum of the elements of the pie chart must add up to the whole. Return to the nonformula display (**Tools | Options** and uncheck **Formulas** on the **View** tab), and block select cells B19 to B21. Click on the Chart Wizard icon on the main toolbar (or click on **Insert | Chart**). You should see the first window of the Chart Wizard. Select the Pie Chart as in Figure 14-12. Click on **Next**.

The second step (Figure 14-13) gives us an idea of what we are going to see. Since we have already predefined the data block, no further entry is needed so we click on Next.

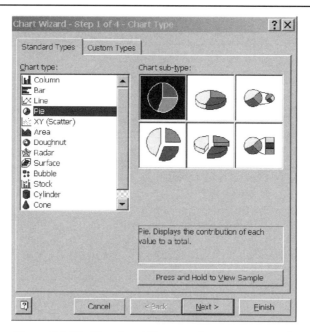

Figure 14-12: Step 1 – We are going to create a pie chart.

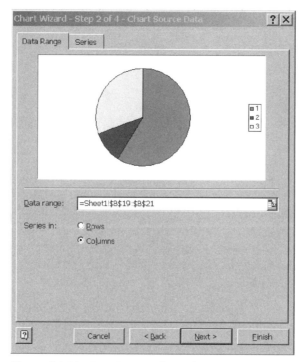

Figure 14-13: Step 2 – We have the option to modify the selected data series.

In the third step, we can add a title on the **Titles** tab. Click on the **Legend** tab and deselect the **Show** legend option. Then click on **Next**.

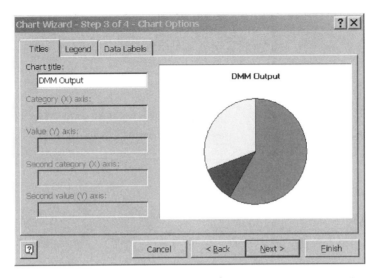

Figure 14-14: Step 3 – Some cosmetic changes.

Finally, we need to decide where the chart will reside. I decided that it should be on Sheet1 (see Figure 14-15). Then click on **Finish**.

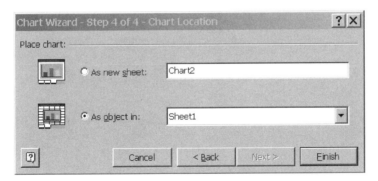

Figure 14-15: Step 4 – We place the chart on Sheet1.

We maneuver the chart to a convenient place on the sheet as in Figure 14-16.

Now we want to get rid of the largest area of the chart. Click on this area until it is selected. You may need to click twice. Then right-click and select **Format Data Point** from the pop-up menu. Under the **Patterns** tab (Figure 14-17), we set the selection to no border and no Area fill effect.

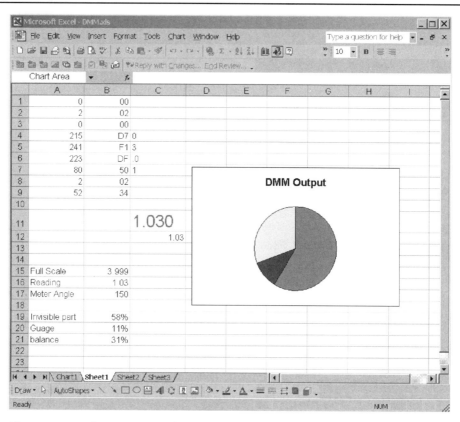

Figure 14-16: Chart is now resident on Sheet1.

Figure 14-17: Making the largest area invisible.

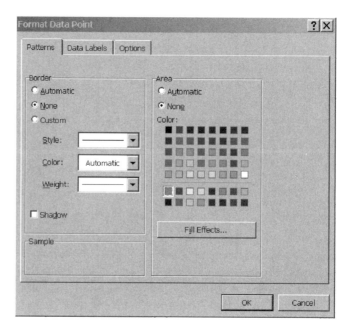

Next we click on the **Options** tab. We can see that the largest slice has already vanished. Modify the **Angle of the first slice** to 75 degrees and deselect the **Vary colors by slice** option. The result is Figure 14-18. Click on **OK**.

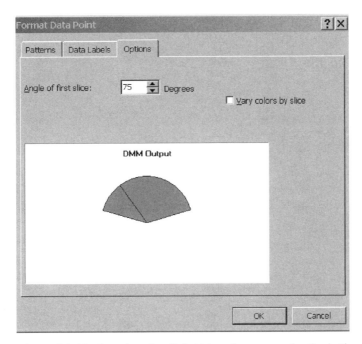

Figure 14-18: Rotating the dial. Using the same color for both slices makes the separating line appear like a needle.

We can change the background color to anything. I chose to go with white. Click on both segments in turn so they are selected (again you may need to click twice), and right-click. Select **Format Data Point** and then select the white color in the **Patterns** tab. As an afterthought, I suppose you could change the background color if an alarm exists.

Before we decide to run the data acquisition, I need to mention something that will save you some confusion. For some reason that I do not understand, when the chart is placed on Sheet1, the code in the PollDMM procedure:

```
For i = 0 To 8
  Cells(i + 1, 1) = iDMMMessage(i)
Next i
```

will not run without generating an error. If the chart is on another sheet there is no problem, although we would need to get rid of the *Sheets("Sheet1").Select* statement that occurs just prior to the above code. If it is left in, viewing the chart becomes quite irritating with the continual return of Sheet1. Of course, this would necessitate changing the *Cells(i+1,1)* to include a reference to Sheet1. I actually take this approach later in the example, but for the moment, let it ride.

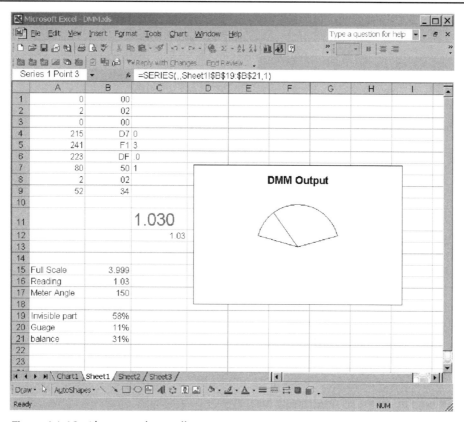

Figure 14-19: Almost ready to roll.

I also tried experimenting with concatenation of strings adding the iteration number "i" to a string containing "A" to allow a slightly less elegant approach (as in Range("Ax") where x is the string value of i), but that caused the same problem. However, if we take the brute force approach of writing each cell directly, it works well. That section of PollDMM becomes:

```
If bBlocReceived = True Then
    'block has been received.
    'write them to sheet1

    Sheets("Sheet1").Select

    Range("a1").Value = iDMMMessage(0)
    Range("a2").Value = iDMMMessage(1)
    Range("a3").Value = iDMMMessage(2)
    Range("a4").Value = iDMMMessage(3)
    Range("a5").Value = iDMMMessage(4)
    Range("a6").Value = iDMMMessage(5)
```

Range("a7").Value = iDMMMessage(6)
Range("a8").Value = iDMMMessage(7)
Range("a9").Value = iDMMMessage(8)
  bBlocReceived = False
  'clear block and start looking again
End If

Inelegant it may be, but there is something to be said for a method that works! This has been stored as *DMM2.xls*.

Zone Identification

Before you modify the procedures for this new approach, please consider the information in "In Parenthesis: Custom Toolbar Limitation."

If we like, we can add an outer ring to the meter with indication for certain ranges. (Take a look at Figure 14-24 for our objective.) In order to do this, we must clear the existing chart since we need to start again. We modify the angle of the meter to 180 degrees just to simplify matters and we create a second table in cells D19 to D22. The cells in D19 to D22 represent the different ranges that should appear while the cells in E19 to E22 define text to appear in the ranges.

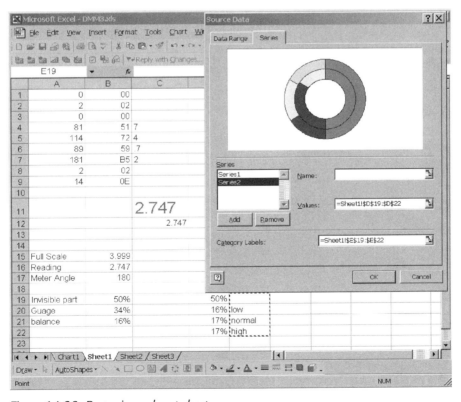

Figure 14-20: Preparing a donut chart.

After highlighting cells B19 to B21, we click on the Chart Wizard creating a Donut chart and "whiz" through only stopping to uncheck the Legend Display option. Next, click on the chart so that the chart as a whole is selected. Right-click on it and select Source Data. In the dialog that appears (Figure 14-20), under the **Series** tab, click on **Add** and define the new series for D19 through D22. Then click in the Category Labels bar and select the range E19 to E22. It is important that we do this here. If we try to add this after we rotate the figure, the category labels will get confused. Click on **OK**.

Once the Chart Wizard has been completed, click on the donut so that the outer circle is selected (indicated by four little squares around the circumference and not a square containing the circle). Right-click and select **Format Data Series**. Under the **Data Labels** tab, select **Category Name** as in Figure 14-21.

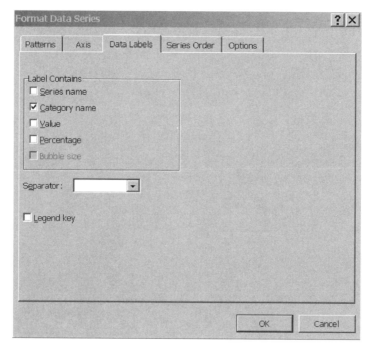

Figure 14-21: Placing category names within the outer band.

Under the **Options** tab, modify the settings as in Figure 14-22.

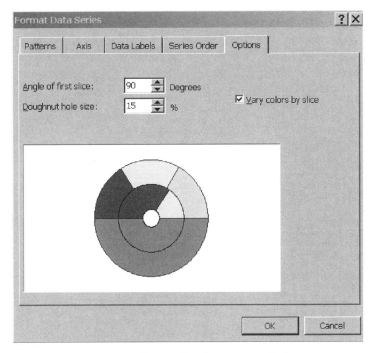

Figure 14-22: Rotating chart and reducing donut hole size.

Figure 14-23 is the result. By selecting the lower segments as before, we can make them invisible. We also change the color behind the "needle" to white and the colors of the zone by selecting each segment in turn, right-clicking, selecting **Format Data Point** and then modifying the settings under the **Patterns** tab.

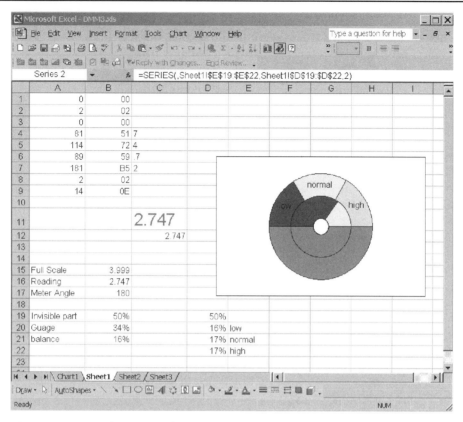

Figure 14-23: Initial setup complete. Note the Category labels. Each can be selected and reformatted on an individual basis.

We are almost there. We can add the contents of any cell as the title of the chart, although it is a little convoluted. Select the chart as a whole, and then right-click and select **Chart Options**. Under the **Titles** tab, enter anything and click on **OK**.

Now click on the newly entered title so that it is selected, click in the formula bar of the main toolbar and enter "=" and the location of the desired cell. You can click on that cell or enter the cell location in longhand. As you can see in Figure 14-24, this can be a dynamic reading. In this case we have placed the DVM reading above the analog meter. This is stored as *DMM3.xls*.

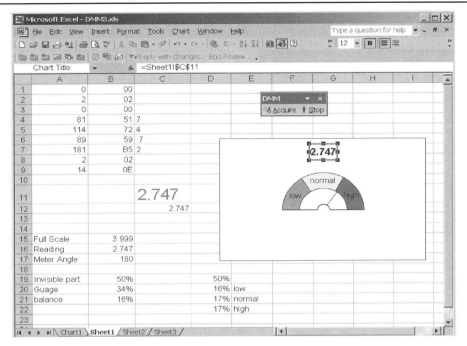

Figure 14-24: Using a donut chart with two ranges to give a VU meter effect. Note the title reference in the formula at the top.

Data Plot—Chart Recorder

Before you modify the procedures for this new approach, please consider the information in "In Parenthesis: Custom Toolbar Limitation."

Let's get rid of the donut/pie charts and consider how to create a graphical representation of the changes on the DMM as a continuously updating chart. In order to do that, we need to take a reading periodically. If we add a value on Sheet1 in cell B14 (named *SampleTime*), we use this value to determine when to take a reading. We then add a variable *nPeriodic* in the declarations and we modify the heart of the PollDMM procedure as follows:

```
'now to format port to read 9 bytes
bSeekingSync = False
UserForm1.MSComm1.RThreshold = 9
'interrupt after 9 bytes
UserForm1.MSComm1.InputLen = 0
'transfer all received bytes when read
iCounter = 0
'initialize
nPeriodic = Timer
While bEndFlag = False
    DoEvents
```

```
If bBlocReceived = True Then
    'block has been received.
    'write them to sheet1
    For i = 0 To 8
        Worksheets(1).Range("A1").Cells(i + 1, 1) = iDMMMessage(i)
    Next i
    bBlocReceived = False
    'clear block and start looking again

    If Timer > nPeriodic + Range("SampleTime").Value Then
        Worksheets(2).Range("A1").Cells(iCounter + 2, 1) = iCounter * Range("SampleTime").Value
        Worksheets(2).Range("A1").Cells(iCounter + 2, 2) = Worksheets(1).Range("c12").Value
        nPeriodic = Timer
        iCounter = iCounter + 1
    End If
End If
Wend
```

Note that instead of selecting Sheet1 and then writing the received data, the storage has the sheet name explicitly included in the instruction. As a result, we can change worksheets without being returned to Sheet1 every time a reading is taken. When the sample time expires (as set on the *nPeriodic* variable), the decoded value in cell C12 on Sheet1 is stored on Sheet2 along with the accumulated time. The first reading is stored in A2 and B2, and then in sequence in A3, B3 and so on. Figure 14-25 will give you the idea.

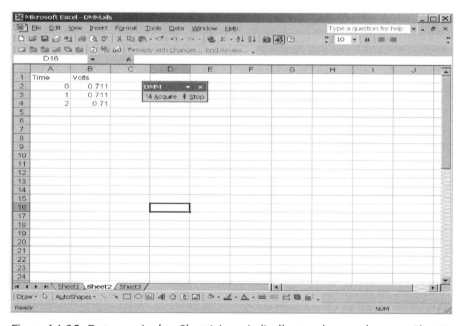

Figure 14-25: Data acquired to Sheet1 is periodically stored as a column on Sheet2. Note the DMM toolbar is visible on any sheet.

Now we want to create a chart. Highlight cells A2 to B5 and click on the Chart icon to bring up the wizard, or as before you can access it through the menus. You should see Figure 14-26.

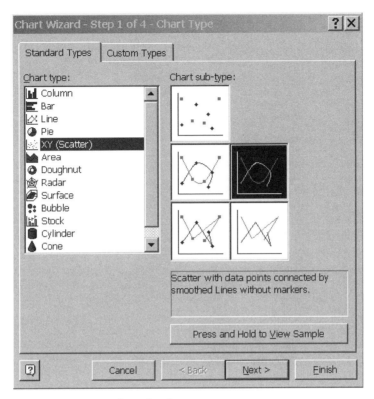

Figure 14-26: Specifying the chart type.

Select the **XY (Scatter)** type and select the sub-type as shown. Click **Next** twice, and on the third dialog (Figure 14-27) enter the titles and modify the appearance of the chart to suit your tastes.

Click on **Next**, and in the fourth step place the chart on Sheet2. The chart should appear looking much like the preview in Figure 14-27. However, this chart is static. The input range is fixed and we would really like the chart to expand or contract to include all the readings, no matter how many there are. Here's how we achieve that.

Click on **Insert** | **Name** | **Define**. As shown in Figure 14-28, type the word *Time* in the **Names in workbook** bar and in the **Refers to:** bar enter the formula:

=OFFSET(Sheet2!$A$2,0,0,COUNTA(Sheet2!$A:$A)-1)

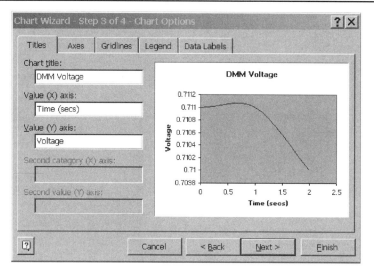

Figure 14-27: Cosmetic changes to the chart.

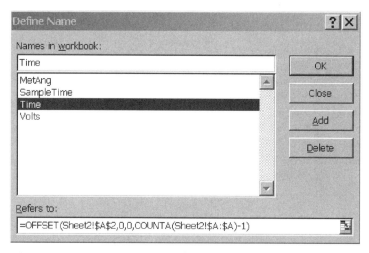

Figure 14-28: Creating a name that automatically extends to cover all the entries in a column.

Similarly create a *Volts* range and include the formula:

=OFFSET(Sheet2!$B$2,0,0,COUNTA(Sheet2!$B:$B)-1)

In Parenthesis: *OFFSET*

The OFFSET function returns a reference to a range of cells. The format is:

OFFSET(reference,rows,cols,height,width)

The reference is the base cell or range of cells. The rows parameter defines how many *rows* away from the base to start, and the same is true for the *columns* with obvious variations. The *height* is how many rows to include in the range. Ditto for *width* in the second dimension. If *height* or *width* are omitted, they will default to the setting of the *reference* cell(s).

Essentially this process allows the named ranges to change dynamically to match the growth of the data. This is as a result of the use of the COUNTA function within the Name formula.

In Parenthesis: *COUNTA/COUNT/DCOUNT/DCOUNTA/COUNTBLANK*

The COUNTA function counts the number of nonempty cells in a range. It differs from the COUNT function in that COUNT only includes numbers in the returned value.

DCOUNT and DCOUNTA count the number of cells in a range that meet specific criteria.

COUNTBLANK does the opposite—it counts the number of empty cells.

Now click on the actual curve in the chart and modify the formula bar in the main menu to read as follows (Figure 14-29):

=*SERIES(,DMM.xls!Time,DMM.xls!Volts,1)*

The selections for the range are now replaced by names, and as we have seen, the names we have chosen will dynamically adjust.

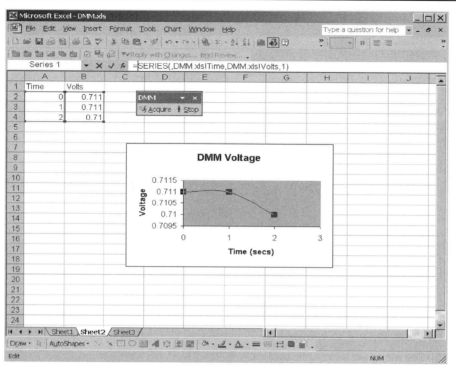

Figure 14-29: Redefining the series used on the chart.

In Parenthesis: *SERIES Function*

Every chart employs a series formula to define the curve. It is not accessible as an entry in a cell in a worksheet, but can be edited as in this example. The format is

SERIES(name, category_labels, values, sequence, sizes)

the *name* in entry is optional and refers to a cell that holds the Series Name used in the legend.

The *category_labels* refers to a range of cells that contain category labels that will be used in the chart. It is easier to get a feel for what this means when you realize that on an XY chart, it is the range used for the X values. Its use is optional and if omitted, Excel will resort to integer numbers starting from 1.

values is the variable that is described in the chart. On XY charts, it is the Y value.

The plotting *sequence* with multiple series determines which series is plotted first, but it must be included even with single series charts.

sizes defines the a range that is used to determine the size of bubbles on a bubble chart.

Since Excel automatically scales the axes, all that is left is to see how the model performs. Ensure that you have clicked away from all the objects so that nothing is selected before clicking on **Acquire**. I charged a 1000 µF capacitor to 1.8 Volts and then discharged it through a 12KΩ resistor. You can see the graphical effect as the discharge starts (Figure 14-30), and then at a later stage as shown in Figure 14-31.

Figure 14-30: Capacitor discharge in the early stages.

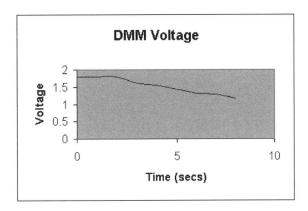

Figure 14-31: Capacitor discharge curve several seconds later.

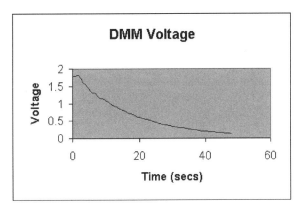

Wait! There's more! Click on the curve so that it is selected, right-click and then click on the **Add Trendline** option from the pop-up menu. Figure 14-30 will be the resulting dialog.

Since we know that a discharging capacitor is exponential, click on the **Exponential** icon. Under the **Options** tab (or at a later stage, right-click on the trendline), add the choice to **Display equation on chart** (Figure 14-33).

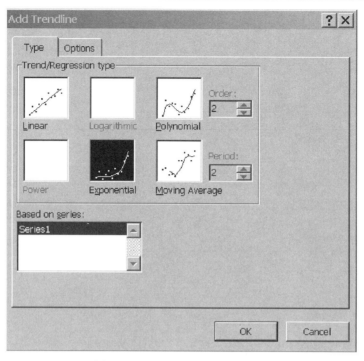

Figure 14-32: Adding an exponential trendline to the chart.

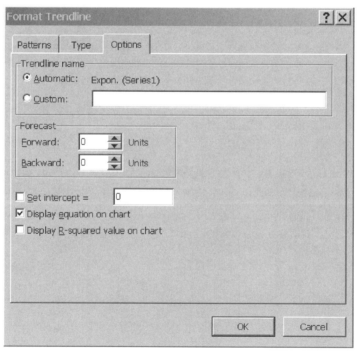

Figure 14-33: Placing equation on the chart.

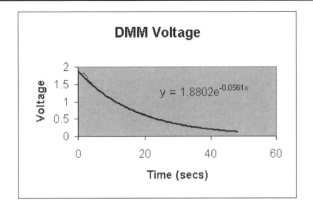

Figure 14-34: And the result is shown on the chart (after a little positioning).

The result (Figure 14-34) is close to the theoretical: $y = 1.8e^{-0.083t}$. This model is on the CD-ROM as *DMM.xls*.

Food For Thought

Since we now know how to manipulate the length of a series on a chart, it should come as no surprise that it is fairly easy to display the last N readings. Let's assume that we create a cell on the worksheet named *NumberOfReadings*. The OFFSET formula we used before could be modified to:

=OFFSET(Sheet2!$B$2,0,0,COUNTA(Sheet2!$B:$B)- NumberOfReadings -1,0, NumberOfReadings,1)

The Excel charting feature does not normally use values that are in hidden rows. This can be changed in **Tools | Options** under the **Chart** tab, but this could be used to our advantage to plot every Nth reading only. Let's assume we only want every 5th reading. On Sheet2 in column C (now on *DMM4.xls*), enter the title *Decision* in cell C1 (just for aesthetics). In cell C2, enter the formula

=MOD(a2,5)

where the 5 derives from the number of readings. Copy this cell starting from C3 down for all the corresponding cells in columns A and B that have data in them. The MOD function merely returns the remainder from a division by 5. Now if we hide every row where the remainder is not zero, then we will have achieved our aim. There is a very quick way to do this. First, click on any cell in column C. Next, click on **Data | Filter | Autofilter**.

Notice how a spinner control is added to columns A, B and C (as in Figure 14-35). Clicking on the 1 in column C and selecting 0 removes 80% of the data. It's like magic! This could easily be added to any procedure as a final step. Removing the Autofilter results is merely a case of repeating the selection process.

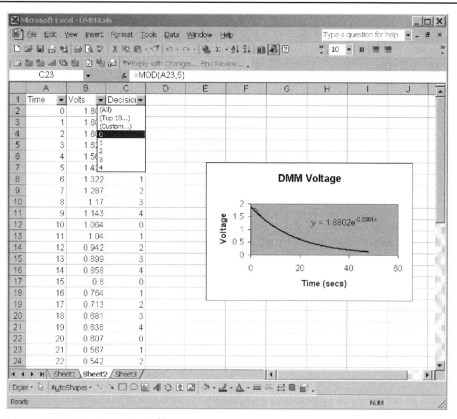

Figure 14-35: Using an Autofilter.

EXAMPLE 15

Vernier Caliper Interface

Model Description

The vernier caliper is very definitely a device that belongs in the mechanical engineering realm, so you may be puzzled why it appears as a topic in a book on Excel in electrical engineering. You will soon discover that it provides the perfect foil for a description on how to use the PC parallel port to interface to the real world, as well as a discussion on statistics.

Mitutoyo Corporation produces a whole host of products to measure linear displacement. Several of these actually do have PC interfaces. The CD-6"C (Code Number 500-171) is a 6-inch vernier Caliper with an electronic interface. As far as I can tell from the web site (www.mitutoyo.com), there are several methods to get data from the caliper into Excel, but all require an extra hardware interface. One method uses a hardware interface to simulate keyboard strokes. Another converts the output to RS-232 and then requires additional software to place the data in Excel. If you are prepared to make a very simple hardware interface consisting of one transistor and four resistors, the capability to read the caliper data into Excel and the ensuing statistical data analysis is a cinch.

Figure 15-1: Vernier caliper with cable attached.

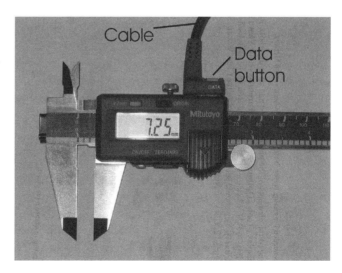

The Mitutoyo vernier caliper has a 5-way connector to provide access to the interface signals. The connections, however, are simply gold plated PCB tracks and you will need the Mitutoyo cable assembly (959149 for a 1-meter cable, 959150 for 2 meters) to make contact with the connector. The price for the cable is reasonable (about $25), and it includes a Data button that can be used to initiate a data transfer as you can see in Figure 15-1.

Pinout

The connector on the other end of the cable has a 2 × 5-way 0.1" header. Figure 15-2 shows the pinout looking at the connector head-on.

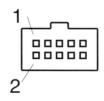

Figure 15-2: Cable pinout. No identification of pin 1 is provided so I simply defined one.

Pin No.	Function
1	Gnd
2	Data
3	Clock
4	Button (to Gnd)
5	!Request

Hardware Interface

There is one signal that goes to the caliper (!Request) from the user interface, and two signals (Clock and Data) from the caliper to the user interface. There is also a ground reference line, and finally a signal that is shorted to the ground when the Data button is pressed. This signal does not connect to the caliper at all. When a data train is initiated by the !Request signal going low, the caliper shifts data out on the Data line with the synchronous separate clock signal. The two outputs each drive an open drain transistor. This allows the use of pull-up resistors to a maximum of 7V for a broad range of interfaces. These are some of the resistors we must add when we construct the hardware interface. The !Request input has its own internal pull-up resistor and should be driven from an open drain/collector configuration. This is the transistor we must build into the hardware interface. Figure 15-3 shows the electrical circuit required.

The printer port connector on the back of the PC does not have a 5V (or any other voltage) as an output. Instead, since the interface draws very little power, we can provide a voltage from one of the data outputs to the pull-up resistors. The rationale is that if it is good enough for a "1" on the output, it should be good enough for the input as well.

Any transistor with a reasonable gain can be used. It should be no problem to saturate the transistor as the collector current is very low.

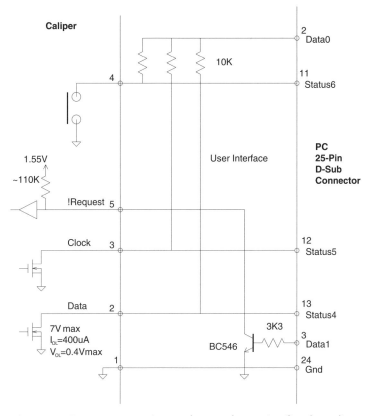

Figure 15-3: Interconnection and user electronics for the caliper interface.

Timing Diagram

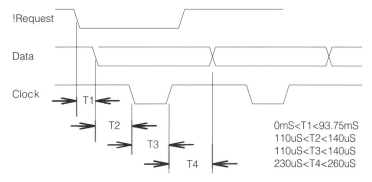

0mS<T1<93.75mS
110uS<T2<140uS
110uS<T3<140uS
230uS<T4<260uS

Figure 15-4: Caliper timing diagram. Note that if !Request is kept low continuously the sequence is repeated about every 90 mS. !Request can be taken high at anytime before the start of the next sequence to prevent a new transmission.

The timing relationship is shown in Figure 15-4 at the bit level. For each reading, the caliper waits for the !Request (the exclamation point indicates that the signal is active low) signal to go low and then transmits 52 bits of information. The information is grouped into 4-bit nibbles as shown in Table 15-1. Each nibble is shifted out with the LSB first.

Output Order	Digit	Name	Description
1	D1	Digit 1	All ones "1111"
2	D2	Digit 2	All ones "1111"
3	D3	Digit 3	All ones "1111"
4	D4	Digit 4	All ones "1111"
5	D5	Sign	+: 0000 –:0001
6	D6	Digit 6	Binary Coded Digit (BCD) Most Significant: =0000 if blank
7	D7	Digit 7	Binary Coded Digit (BCD)
8	D8	Digit 8	Binary Coded Digit (BCD)
9	D9	Digit 9	Binary Coded Digit (BCD)
10	D10	Digit 10	Binary Coded Digit (BCD)
11	D11	Digit 11	Binary Coded Digit (BCD) Least Significant
12	D12	Decimal Point	Indicates how many digits to the right of the decimal point. Can range from 2 (0010) to 5 (0101)
13	D13	Unit	mm: 0000, inch: 0001

Table 15-1: Details of nibble transmission from the caliper.

Installing IO.DLL

In order to access the I/O lines of the PC parallel interface, you need some form of software interface that will bypass the operating system and control the hardware directly. Appendix B discusses the options and their availability in more detail. I opted for the Data Link Library called *IO.DLL*.

With the permission of the author, Fred Bulback, the program is included on the CD-ROM. Copy the file, IO.DLL, to the System or System32 subfolder in the Windows/System folder. Cut and paste the Visual Basic declarations from the Geek Hideout web site to the General Declarations in Module1 in VBA. If you are working with "Caliper.xls", obviously the declarations are already there. I suppose you could just as easily copy the declarations from here for a new application.

PC Parallel Port

The PC parallel port was originally intended for a Centronics printer interface. The addresses were referred to as LPT1 through to LPT3. Each printer port consisted of a series of I/O addresses at one of three base addresses as in Table 15-2. The actual address that your computer will use depends on the age of your machine, but Table 15-2 is the most likely configuration. You can find out what your system is using by going to the System Properties window of your computer's Control Panel. There will be a listing of the port addresses in the Device Manager section.

Table 15-2: Printer port base addresses.

Port	Address
LPT1	0x378
LPT2	0x3BC
LPT3	0x278

As newer parallel port approaches evolved, these interfaces became more flexible, but in its basic form the port consists of an 8-bit latched output port located at the LPT base address called the *Data register*. At LPT +1 there is a 5-bit input register known as the *Status register*. The *Control register* at LPT +2 adds 4 digital outputs. The pinout of the 25-pin parallel port is given in Table 15-3. The original function of the bits is irrelevant to our discussion, but if you would like to know more about this or the other variations of the port, see the excellent book, *Parallel Port Complete*, by Jan Axelson.

D-sub pin	Signal (Register + bit number)	Function	Direction (relative to PC)	Inverted
1	nStrobe	Control bit0	Out	Y
2	D0	Data bit output, bit0	Out	N
3	D1	Data bit output, bit1	Out	N
4	D2	Data bit output, bit2	Out	N
5	D3	Data bit output, bit3	Out	N
6	D4	Data bit output, bit4	Out	N
7	D5	Data bit output, bit5	Out	N
8	D6	Data bit output, bit6	Out	N
9	D7	Data bit output, bit7	Out	N
10	nAck	Status bit6 (may trigger interrupt)	In	N
11	Busy	Status bit 7	In	Y
12	Paper End	Status bit5	In	N
13	Select	Status bit4	In	N
14	nAutoLF	Control bit1	Out	Y
15	nError	Status bit 3	In	N
16	nInit	Control bit2	Out	N
17	nSelectIn	Control bit3	Out	Y
18-25	Gnd			

Table 15-3: Parallel port function and pinout.

First Steps

Open an Excel workbook and go to VBA (**<Alt>** + **<F11>**). Add a module (**Insert | Module**), and copy the Visual Basic declarations (as discussed above) to the General Declarations section of the module. In actual fact, for this example, you will only need the PortOut and PortIn declarations.

The nine-month-old computer that I am using for this development is a 1.8 GHz Celeron with 500 MB of RAM running Windows 2000. I was not sure how fast the I/O switching would be, so I wrote a short VBA program to toggle an output (pin 2 of the port, D0) along with some fictitious operations that would give me some idea of the cycle time. This is the program:

```
Sub test()
    Dim i As Integer
    Dim j As Integer
    Dim k As Integer
    Dim m(25) As Integer

    For i = 1 To 20
      k = 0
      For j = 0 To 8
         Call PortOut(888, 1)
      '888=0x378
          k = k * 2
          k = k Or 1

         Call PortOut(888, 0)
      Next j
      m(i) = k
      m(i) = k * 2
      'DoEvents
    Next i

    End Sub
```

In the worst case, the first cycle took 13.1 µS and all the following cycles improved to 2.9 µS (see Figure 15-5), so reading the output of the caliper should be well within the capability of this computer. I also added a *DoEvents* instruction to see what the effect would be; the performance deteriorated so badly that I didn't even try and measure it. The implication here is that while we are reading data from the caliper, any other process must be frozen until reading is complete.

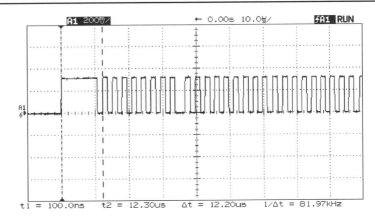

Figure 15-5: Assessing the execution times of my computer. The longest cycle time was always the first as shown here at 12.20 µS.

Actual Interface

After that lengthy prelude, we are ready to begin. The data from the caliper will be read on bit 4 of the Status register and the clock will be on bit 5. Power to the pull up resistors of the interface will be derived from bit 0 of the data register. The !Request signal is derived from bit 1 of the data register inverted by a transistor (see Figure 15-3). The Data pushbutton signal is read in the status register bit 7.

The principle of operation is as follows:

1. Wait for Data pushbutton to go active.

2. Activate !Request signal.

3. In groups of 4 bits, wait for the clock bit to go active, read in bit, and wait for clock to go inactive combining the 4 bits to generate a digit.

4. Repeat for the remaining twelve digits.

5. Deactivate !Request.

6. Create number from the readings and place in worksheet.

7. Wait for the Data pushbutton to go inactive.

Acquiring Data

It seems to me that a natural mode of operation in measuring a series of readings (for statistical analysis) would consist of starting the sequence with a Start button on the Excel interface, acquiring caliper data every time the Data button is pressed, and clicking on a Stop button on the Excel interface to terminate the acquisition and allow data analysis.

Let's deal with the Stop function first. We will use a single global variable that is set when the stop button is clicked, and cleared when acquisition starts. The procedure that will be called when the Stop button is clicked is:

```
Sub StopRequest()
    bStopRequest = True
End Sub
```

and this will be placed in Module1, which should still be active from the earler execution time tests.

When the workbook is first opened, we would like the pull-up resistors in the interface powered and the Request line to the caliper disabled, so in the Worksheet Activate event we call a call to the procedure SetupPort which is also in module 1.

```
Sub SetupPort()
    Call PortOut(888, 1)
    'ensure power is applied to pull up resistors
    'and the !request line is inactive
    bStopRequest = False
End Sub
```

The heart of the process is invoked from a Start button which will run the Capture procedure. This is quite lengthy, so let's discuss it in parts. Obviously, the code exists in "Caliper. xls". Aside from the memory declarations, the procedure is initialized as follows:

```
'first clear contents
Sheets("Sheet1").Select
Range("A:A,B:B").Select
Range("B1").Activate
Selection.ClearContents
Cells(1, 4) = ""
Cells(1, 5) = ""

iCellPoint = 0
'intialize pointer
Cells(1, 4) = "REC"

bStopRequest = False
'initiating the condition
```

The results of the readings are placed in columns A and B, so the first thing to do is to clear all the previous results. Also, when data is being acquired, cell D1 contains a red "REC" (the text in the cell is formatted during worksheet setup) and cell E1 contains the total number of readings acquired since initiation. Actually, it is more of a pointer that is incremented every time a reading is taken.

The continuous loop is achieved through the "*while 1*" statement at the start of the following code. The corresponding *wend* comes right at the end of the procedure and is not shown here. The Data button on the caliper is connected to bit 7 of the Status register of the parallel port. This input is inverted so when the button is pressed connecting the signal to ground, the software will see this as a digital one. While it is scanning for this input, the procedure allows other events to occur, so that if the Stop button is clicked, the bStopRequest flag will be set and the procedure can be terminated.

```
While 1
'do forever

  j = 0
  While j = 0
  'waiting for Data switch to be pressed
  'it goes low, but this input is inverted
    DoEvents
    j = PortIn(889) And 128
    If bStopRequest = True Then
      Cells(1, 4) = ""
      'remove REC symbol
      Exit Sub
    End If
  Wend
```

Once the Data button is seen, the procedure reads the input without allowing for other system events. Errors in the read process can happen, which is not surprising since the signals are unbuffered and are at very low current and low voltage levels. Although the caliper serial data protocol does not include checksums, it is possible to add some checks to detect the errors. When an error is detected, the read process is reinitiated using the inelegant *go to* approach. I couldn't think of anything simpler. *Retry:* is the *go to* address.

The !Request signal is first disabled, and then enabled after a period of 100 mS to allow the caliper to reset. Bits are then read in nibbles as the clock is detected. On occasion, the read sequence can go out of phase and the procedure will lock-up waiting for a clock pulse and none arrives. I added a simple counter (since the timer will not work without *DoEvents*) and found a suitable value by trial and error. This value is likely to vary between processors. If the value is too large, then it may take some time to detect and then clear the error. If the value is too small, the process will be endlessly repeated because of false timeouts. At any rate, each time the loop waiting for the clock pulse is executed, the counter iErrorDetect is incremented. When the counter gets too large, the acquisition process is restarted. As a quick note, it is not possible to use the *Exit* statement within a *while/wend* loop. It can only be used in the alternative construct of *do/loop while* loop.

Once all the bits have been read, the !Request line is deactivated.

```
Retry:
    Call PortOut(888, 1)
    'ensure system is off for retry
    nTimerSave = Timer
    While Timer < nTimerSave + 0.1
```

```
'wait for debounce
Wend

Call PortOut(888, 3)
'REQUEST signal low (after transistor)

For i = 0 To 12
   '13 digits
   iDigit = 0
   For k = 0 To 3
   '4 bit per digit
      j = 1
      'flag for while statement
      iErrorDetect = 0
      Do
         'waiting for input ot go low
         j = PortIn(889) And 32
         iErrorDetect = iErrorDetect + 1
         If iErrorDetect > 50000 Then
            Exit Do
         End If
      Loop While j <> 0
      iX = PortIn(889) And 16
      'bitwise and
      iDigit = iDigit / 2
      'shift right
      If iX <> 0 Then
         iDigit = iDigit Or 8
         'oring on msb
      End If
      j = 0
      'flag for while statement
      While j = 0
         'waiting for input ot go high
         j = PortIn(889) And 32
      Wend
   Next k
   iDigitArray(i) = iDigit
Next i

Call PortOut(888, 1)
'REQUEST signal low (after transistor)
```

With the 13 nibbles stored in *iDigitArray*, a check is made for suitable values in some of the bytes. If an error is detected, the reading is discarded and another one taken.

```
'error check
If iDigitArray(0) <> 15 Or _
iDigitArray(1) <> 15 Or _
iDigitArray(2) <> 15 Or _
iDigitArray(3) <> 15 Or _
iDigitArray(11) > 5 Or _
iDigitArray(11) < 2 Then GoTo Retry
```

The procedure then takes the digits that have been read in and formats them in a string adding the decimal point in the correct spot. Then the result is converted to a number and stored at the desired cell in the worksheet.

```
i = iDigitArray(11)
'fetching how many digits there are to the right of the dp
sNumber = iDigitArray(10)
For i = 1 To iDigitArray(11) - 1
    sNumber = iDigitArray(10 - i) & sNumber
Next i
sNumber = "." & sNumber
For i = 10 - iDigitArray(11) To 5 Step -1
    sNumber = iDigitArray(i) & sNumber
Next i

Cells(iCellPoint + 2, 1) = iCellPoint
Cells(iCellPoint + 2, 2) = sNumber
Cells(1, 5) = iCellPoint

iCellPoint = iCellPoint + 1
```

The procedure takes less time to execute than it does to explain, and it is quite likely that the Data button is still activated when the procedure iteration is complete. As a result, there needs to be some code to wait for the button to be released.

```
j = 128
While j <> 0
'waiting for Data switch to be pressed
'it goes low, but this input is inverted
    DoEvents
    j = PortIn(889) And 128
    If bStopRequest = True Then
        Cells(1, 4) = ""
        Exit Sub

    End If
Wend

nTimerSave = Timer
While Timer < nTimerSave + 0.2
'wait for debounce
Wend
```

Having created the code, we add two command buttons using the Forms control as shown in Figure 15-6. Link them to the corresponding procedures, and then format the text in cell D1 to red, so that the REC will appear in red.

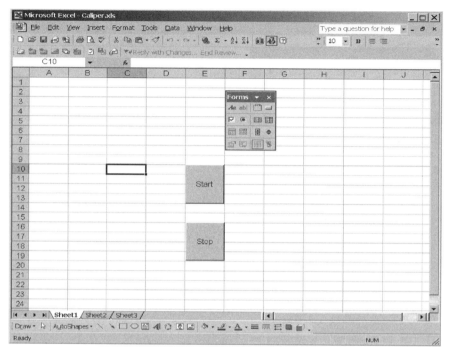

Figure 15-6: Placing two command buttons from the Forms toolbox.

Figure 15-7 shows the acquisition of data from the caliper in operation.

Adding Sound

As each reading is stored, the incremented count appears in cell E1, but it would be nice to give the operator some audio indication that the data has been stored so that they do not have to look at the screen at all. Earlier versions of Excel had a function that generated some kind of sound, but that feature has been removed. There is a "beep" instruction in VBA, but if your computer is relatively modern, it likely does not have the internal speaker. We cannot use the message box because it will require operator interaction to close it for the next reading and that defeats the hands-free approach. The only route available to us is to play a ".wav" file. This is how we do it. First, place the following declaration in the General Declarations area of Module1.

```
Private Declare Function mciExecute _
Lib "winmm.dll" ( _
    ByVal lpstrCommand As String _
) As Long
```

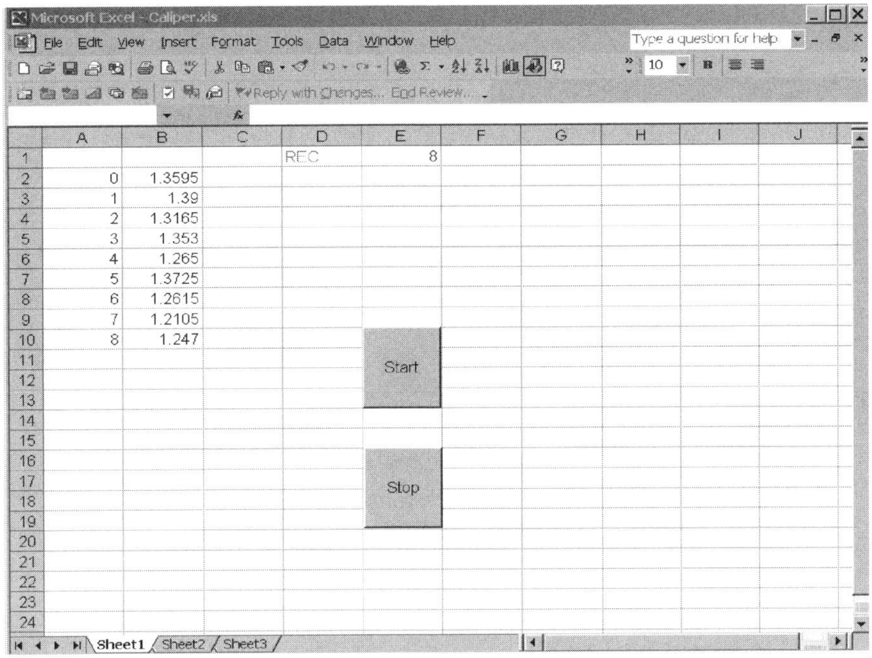

Figure 15-7: Acquiring data. Each reading is added in the next row forming a column of a pair of numbers. Note the data is still being acquired as seen by the "REC" in cell D1.

Remember that the *<space>_* is how a line continuation is achieved in VBA.

In the location where the process has been successfully concluded, we insert the line

$x = mciExecute("play\ c:\backslash winnt\backslash media\backslash ding.WAV")$

It appears just prior to the condition waiting for the Data button to be released. The *DoEvents* in the following while loop allows the sound to be played. Obviously almost any .wav file can be played, but it would probably be sensible to choose a shorter one in this case.

Thoughts on Improvement

Even with the error-checking, some erroneous readings still sneak through. These readings are way off the mark and I suppose as part of the check, we could make a comparison to a nominal value and if it differed by more than say 40%, we could invoke a new reading.

Aside from the printer port, it is possible to buy expansion boards for the internal bus of the computer. Most times, these devices rely on some implementation of the flexible Intel 8255 parallel port adapter. The manufacturer of the card will either provide a driver or the I/O address for the port and you can use the same techniques shown here to achieve greater flexibility (in terms of number of I/O lines and directionality) for more complex projects.

Statistics

Excel has an extensive array of functions for statistical analysis. Most are installed with the Analysis Toolpak. If you have not done this, you should install the add-in as described in the introduction. I am far from being an expert in this field, so I merely want to highlight some of the functions that you can use.

In an attempt to make this example vaguely electronic, I measured the contents of three tubes of 28-pin integrated circuits (27C512s if you must know) with the results shown in Figure 15-10. It is simple enough to generate the average, cell D3 contains the formula

=AVERAGE(B2:B40)

which does not need further explanation. In a similar manner, the standard deviation can be easily calculated and the cell D6 has the formula:

=STDEV(B2:B40)

to quickly calculate it. Finding the minimum and maximum readings is simple with the =*min* and =*max* functions.

Let's create a frequency distribution for these readings. The quickest and easiest way is to use the Data Analysis tool. Click on **Tools | Data Analysis**, and then select **Histogram** and **OK**. You should see the resulting dialog in Figure 15-8.

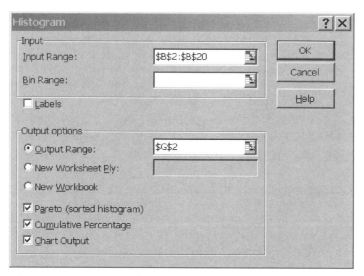

Figure 15-8: Creating a frequency distribution histogram.

The frequency distribution is grouped into bins. You can enter a series of values (not necessarily equally spaced) in a range in the worksheet and enter in the **Bin Range** bar, or you can let the feature do the grouping automatically by leaving the entry blank. Select the options that you want, and Figure 15-9 is the result. You can massage the chart's appearance to your heart's content.

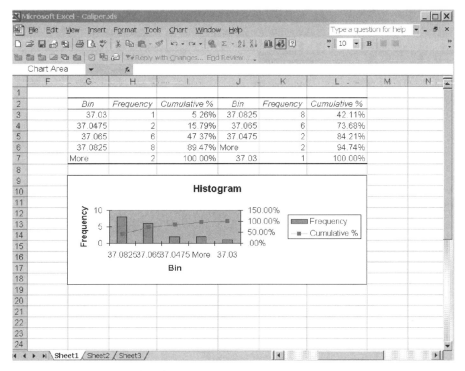

Figure 15-9: Frequency distribution output.

There is an alternative approach that is a little more complex, but may yield more flexibility. At any rate, it allows us to see a few more Excel functions in action. The first item on the agenda is to create the bins, which are equally spaced ranges between the maximum and minimum readings. Five bins would seem reasonable for this small spread of data. Let's create a table of the bin ranges automatically.

Select cells G2 to G6 (a total of 5 cells) and then click in the formula toolbar with the cells still selected. Type in the formula:

$=MIN(B2:B40)+(ROW(INDIRECT("1:5"))*(MAX(B2:B40)-MIN(B2:B40))/5)$

and instead of pressing **<Enter>**, we enter an array formula by pressing **<Ctrl>** + **<Shift>** + **<Enter>**. (See Appendix A for a discussion on array formulas.) The result is seen in Figure 15-10. Note that if you want more bins, the "5" has to be changed in two places in the formula and the number of cells selected must also be changed.

Highlight cells H2 to H6 and in the formula bar enter the formula:

$=FREQUENCY(B2:B20,G2:G7)$

(yes G7! see the FREQUENCY sidebar) followed by **<Ctrl>** + **<Shift>** + **<Enter>** to enter the array formula. The distribution now appears in column H.

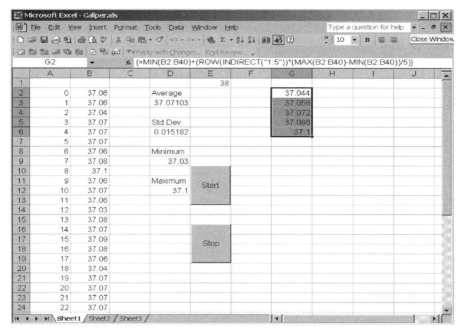

Figure 15-10: Simple statistics applied to the acquired data together with bin creation.

In Parenthesis: *FREQUENCY*

It is possible to determine the number of different values that occur within a range of numbers. The granularity of the range of numbers is expressed as bins each covering a subset of values within the range. The FREQUENCY function returns an array of numbers, and as a result, must be entered as an array formula. The syntax is:

FREQUENCY (data_array,bins_array)

where the data array is the array of measurements where the frequencies are to be measured. For no readings, an array of zeroes is returned.

The bins_array is an array that contains the upper value of each range of the bin. The lower value is defined by the upper value of the previous bin. Since this is in a tabulated form, uneven ranges can be created.

This function returns one additional array element more than the bins_array. It is the value of number of readings above the last value of the bins.

We can also try and see how our readings compare to a normal distribution. In cell H2, enter the formula:

=NORMDIST(G2,mean,std_dev,FALSE)

and copy it to cells C3 to G7.

In Parenthesis: *NORMDIST*

For a given average and standard deviation, this function will return the normal distribution at a given point of the population (the x-axis). The syntax is:

NORMDIST(x,mean,standard_deviation,cumulative)

x is the point at which the distribution will be evaluated.

mean is the average of the function

standard deviation needs no explanation.

cumulative is a logic value. If it is set to TRUE, the cumulative distribution is found. FALSE returns the probability mass.

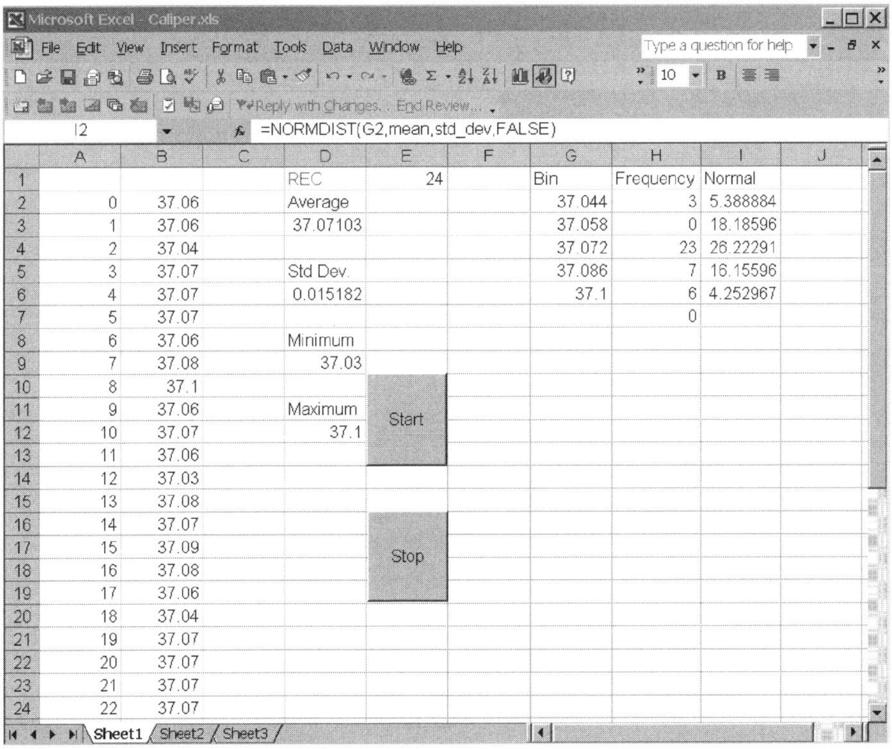

	A	B	C	D	E	F	G	H	I	J
1				REC	24		Bin	Frequency	Normal	
2	0	37.06		Average			37.044	3	5.388884	
3	1	37.06		37.07103			37.058	0	18.18596	
4	2	37.04					37.072	23	26.22291	
5	3	37.07		Std Dev.			37.086	7	16.15596	
6	4	37.07		0.015182			37.1	6	4.252967	
7	5	37.07						0		
8	6	37.06		Minimum						
9	7	37.08		37.03						
10	8	37.1								
11	9	37.06		Maximum						
12	10	37.07		37.1						
13	11	37.06								
14	12	37.03								
15	13	37.08								
16	14	37.07								
17	15	37.09								
18	16	37.08								
19	17	37.06								
20	18	37.04								
21	19	37.07								
22	20	37.07								
23	21	37.07								
24	22	37.07								

Figure 15-11: Results of the statistical analysis of the data in B2:B40.

Figure 15-11 is the result of our efforts to date. The next logical step is to depict this on a chart. Select the Chart Wizard by clicking on the icon, or from the **Insert | Charts** menu. As in Figure 15-12, select the standard line type chart. Click on **Next**.

Figure 15-12: Choosing a chart type.

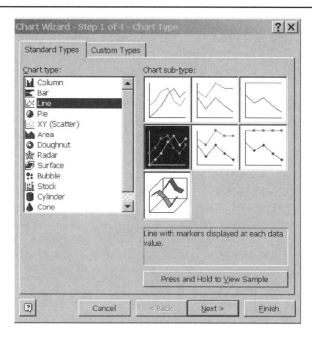

The next step defines the source data. Select the Data range as cells H1 to I6, which includes the titles in the selection as shown in Figure 15-13. Excel will automatically include the text in the first row as the names for the data used in the chart.

Figure 15-13: Establishing the source data for the y-axis.

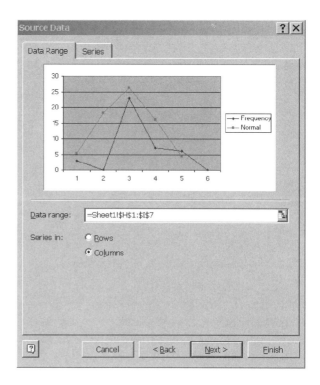

Click on the **Series** tab and enter the cells associated with the bin values in the **Category (X) Data Labels** box as in Figure 15-14.

Figure 15-14: Establishing the source data for the x-axis.

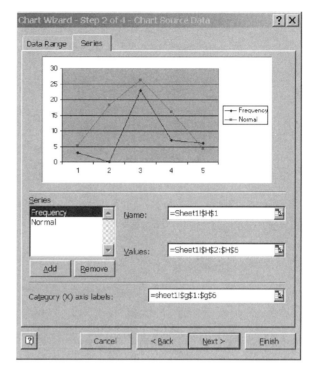

The third step is for cosmetic enhancements, and we will ignore them for the moment. In the fourth step, we place the chart on Sheet1. The result is shown in Figure 15-15.

Figure 15-15: Chart output of the distribution frequency, showing the actual data selected.

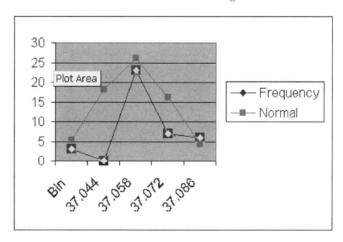

It might improve the appearance of the chart if we change the graph type of the actual readings to columns. Excel will allow us to mix different chart types. To do this we need to click on the associated curve as in Figure 15-15, right-click on it and select **Chart Type**. From the

Chart Type dialog, select the column type as in Figure 15-16. Click on **OK** and Figure 15-17 is the result.

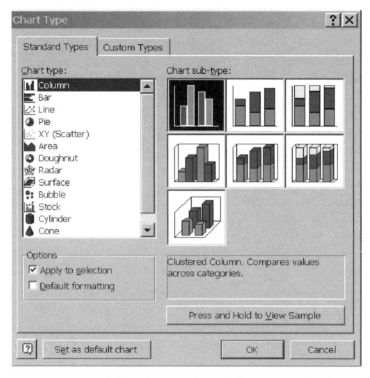

Figure 15-16: Modifying the type of one of the curves on the chart.

Figure 15-17: Chart output of the distribution frequency mixing two different chart types.

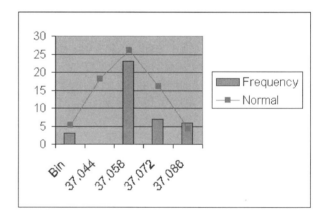

Obviously there are not enough readings, or the bin granularity is not fine enough to show the classic bell shape curve, but I am sure you get the idea.

How's that for convergence? Electrical engineering, mechanical engineering, statistics and computer science—four disciplines in one example!

EXAMPLE 16

Function Generator Interface

Model Description

Our department had just acquired a Stanford Research Systems Model DS345 30-MHz Synthesized Function Generator, and I wanted to use it to generate an adjustable pulse width to test the speed characteristics of a Pulse Width Modulation (PWM) controlled electric motor. Despite the generator's versatility, I found the pushbutton interface of the instrument (Figure 16-1) and its optional DOS-based configuration program did not lend themselves to conveniently realize this function. The idea for this model was born from this need.

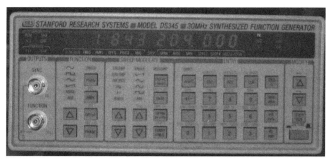

Figure 16-1: The DS345 Synthesized Function Generator.

The DS345 is available with an RS-232 interface. The documentation supplied with the generator is exemplary and is obviously written for exactly this kind of application, although I don't think the designers ever presumed it would be run from Excel.

Generating typical waveforms is easy enough to do from the keyboard, and I am far from an expert on modulation techniques. These two reasons coupled with the desire to develop a simple model have led me to only show a technique to develop custom waveforms along with the ability to skew them.

My rationale for using Excel was that you could create a chart that would reflect the output that you wanted. Since the chart is always based on a tabular input and since we know it is possible to create a chart that expands dynamically to cover the exact amount of data, I thought that it would be easy to create and modify the chart to show complex waveforms. I also allowed for initializing the data in the table to a recognized waveform and then allowing further modification. Once the waveform was created, it could be saved (as a scenario perhaps) and archived for use at some other time. In addition, it would be possible to generate information that is certainly not available normally, like RMS voltage and Crest Factor.

Serial Interface

The serial protocol that the DS345 uses includes 8 data bits, no parity and 2 stop bits. The baud rate is programmable, and I selected 9600 baud. The Function Generator must have the serial port enabled and the baud rate set from the keyboard, and this information is covered in the DS345 user manual. The Function Generator front panel even has a display setting to allow the user a view of the received data, making debugging especially easy. The RS-232 interface connector is the original DB25 format. You will need to use a 25-way adapter to a DB9 connector, and then use a "straight-through" 9-way cable to the serial port on the PC.

The control is mostly achieved with ASCII commands, but the custom waveforms are downloaded in binary. The DS345 has many commands, but they are not really pertinent to our needs here, so I will only describe the commands that I am going to use. If you are going to try this yourself, you will no doubt have a DS345 and the manual that goes with it, so you will have a description of all the possible commands.

The command protocol allows for a series of four ASCII characters, followed by some numbers where additional data is required. Spaces are treated as null characters and the command is terminated by a line feed or carriage return.

*RST is the command that resets the DS345.

FUNC 5 sets the DS345 into the custom waveform mode.

FSMP x determines the granularity of the output waveform. The Function Generator output is driven from a D/A converter. Each reading on the converter is held for a period of time. This time period is expressed as a frequency (the inverse of the time) that is derived from the value 40 MHz/(N) where $0<N<(2^{34}-1)$. As a result, x (used in the command) must be an exact divisor of 40MHz or it will be rounded to the nearest allowable frequency. Based on this range each data point of the wave can be held for an interval of 25 nS to 2.3 mS.

LWDF 0,j allows downloading j (a maximum of 16,300) points in the waveform. Each point in the download data is sent as a 16-bit binary number made up of two 8-bit bytes. The number is limited to between −2047 and +2047. The data is terminated with a checksum, which is the 16-bit addition of all the data words transmitted.

Workbook Open and Close

Before we start adding worksheet controls, let's make sure that when the workbook opens the Function Generator is reset. That will involve initializing the serial port, and sending the reset command.

In a new workbook, invoke VBA (**<Alt> + <F11>** or **Tools** | **Macro** | **Visual Basic Editor**) and then add a module (**Insert** | **Module**) and a user form (**Insert** | **UserForm**). With the User Form active, click on the MSComm icon on the toolbox and then place the control on the User Form (see Example 14 and Appendix B on how to get the MSComm icon). See Figure 16-2.

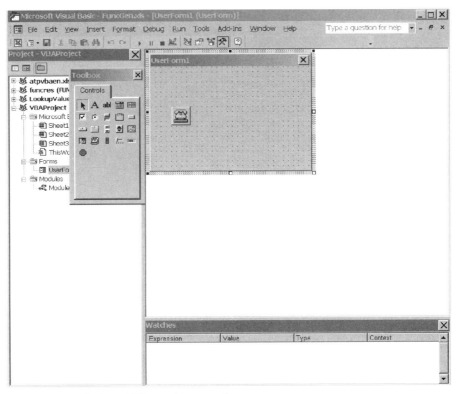

Figure 16-2: Placing MSComm in a user form.

Double-click on the Module1 folder, and in the code window add the following:

```
Sub SerialPortOpen()
'Normally the following commented lines would be used
'and when the model is complete they will be uncommented.
'During development the access to the serial port may go
'out of phase and so, if the port is open we first close
'and then reopen it to prevent false information being read or sent
'
'Created as a procedure so that it only needs to be changed once,
'but can be accessed from anywhere
    'If UserForm1.MSComm1.PortOpen= False Then
        'UserForm1.MSComm1.PortOpen = True
    'End If
'remove when model is complete
```

```
    If UserForm1.MSComm1.PortOpen = False Then
        UserForm1.MSComm1.PortOpen = True
    Else
        UserForm1.MSComm1.PortOpen = False
        UserForm1.MSComm1.PortOpen = True
    End If
End Sub
```

Add a second procedure:

```
Sub Initiate()
    With UserForm1.MSComm1
    'this is just shorthand to save writing
    'UserForm1.MSComm1.xxxx= nnn
    'every time.
        .CommPort = 1
        '.PortOpen = True
        'enabled in a few lines to allow check if open
        .Handshaking = comNone
        'no handshaking required
        .Settings = "9600,N,8,2"
        'set for 9600 baud, no parity, 8 bits, 2 stop bit
        .DTREnable = True
        .RTSEnable = True
    End With
    'add check if port is open
    Call SerialPortOpen
    UserForm1.MSComm1.InputMode = comInputModeText
    'text data

    TransmitBuffer = "*RST" + Chr(13)
    UserForm1.MSComm1.Output = TransmitBuffer

End Sub
```

For a detailed description of the MSComm properties see Example 14. Note the transmission of the 4-character command "*RST" with the concatenation of the carriage return character to complete the message. With the DS345 connected, we can run this procedure to see that it does in fact get reset.

In the Workbook_Open event, add the code:

```
Private Sub Workbook_Open()
    Call Module1.Initiate
End Sub
```

It is good form to close the serial port when the application is closed, so in the workbook deactivate event we add:

```
Private Sub Workbook_Deactivate()
    Call Module1.SerialPortClose
End Sub
```

and create a procedure in Module1 that will close the port:

```
Sub SerialPortClose()
   UserForm1.MSComm1.RThreshold = 0
   'disable interrupts-
   If UserForm1.MSComm1.PortOpen = True Then
      UserForm1.MSComm1.PortOpen = False
   End If
   'close the port
End Sub
```

Adding VBA Controls: Granularity

Return to the Excel workbook from VBA and name Sheet1 to "Controls" and Sheet2 to "Workings". As we have seen elsewhere in this book, there are several ways to introduce controls to Excel. Since I want the ability to dynamically modify some of the controls, I chose to go with the Control Toolbox. The controls will be associated with a particular sheet so we won't have to turn the visibility of the controls on and off, but because these controls are ActiveX controls, the approach will be very similar to VBA controls placed on a form.

In Parenthesis: *Controls in Excel*

We have seen in previous examples that there are four ways to create controls in Excel. Each method has advantages and disadvantages.

Method	Advantages	Disadvantages
Data Validate	Simple data validation. Results and selection in single cell.	No directly associated procedure possible. Only combo box type functionality.
Form Control	Simple to use. Floats above worksheet. Associated with a worksheet.	Lack flexibility and features, for example, cannot be enabled or disabled, cannot dynamically change range of slider control. Difficulty sizing several controls to the same size.
Control Toolbox	ActiveX controls allow for a wide variety. Can be dynamically changed and enabled/disabled. Floats above worksheet. Associated with a worksheet.	More complex setup. May require initialization.
VBA Form	ActiveX controls allow for a wide variety. Can be dynamically changed and enabled/disabled. Floats above worksheet. Not associated with a worksheet.	More complex setup. May require initialization. Requires form visibility control.

In order to start, we need the Control Toolbox visible (**View** | **Toolbars** | **Control Toolbox**). We place a text box with the text "Granularity" on the worksheet, and then format the **TextAlign** property to **center alignment** and the **SpecialEffect** property to a **flat** appearance. Next, we add a combo box control which we will name "Granularity" and link it to cell B1 on the Workings sheet. Make sure that the **Style** property is set to **style 0**. Also, the **MatchEntry** property should be set to 2, or data entry will be somewhat puzzling as the data entry will automatically try to select the value closest to the first character the user enters. You should be looking at something like Figure 16-3.

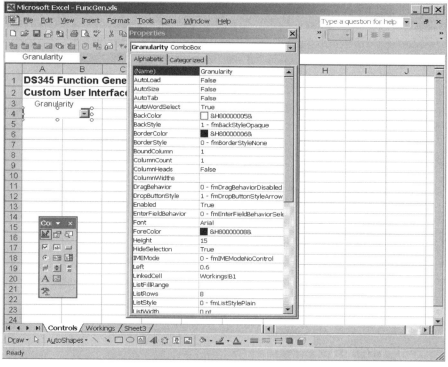

Figure 16-3: Placing a combo box control.

In Parenthesis: *Combo Box Control*

The ActiveX combo box control allows two styles: style **0-fmStyleDropDownCombo** allows the user to enter data or select from the drop-down list. Style **2- fmStyleDrop-DownList** only allows the user to select from the drop-down list. The result appears in a cell identified in the **LinkedCell** property. Each time the entry is changed, a click event is triggered.

The data that appears in the box is also found on the text property of the combo box, but in order to create the drop-down list, the program must initialize the values in VBA, possibly in some initialization event. To do this we use the AddItem method, which takes the format :

   Object.Additem "Text1"
   Object.Additem "Text2"

The VBA help suggests that it is possible to add an index number to allow the user to specify the order of appearance in the drop-down list. From my experience, this variation does not work in the Control Toolbox implementation of the control, and the order of appearance is the order in which the AddItem method is executed.

The ListRows property determines the maximum number of rows shown in the drop-down list.

Before we implement this, let's define "granularity." The output of the function generator is driven by a digital-to-analog converter (DAC). This output is updated periodically, and the more updates there are in the waveform, the smoother the output is (within the limits of the resolution of the DAC), as shown in Figure 16-4. In this context, "granularity" is the number of times the output is updated in one cycle of the waveform.

Figure 16-4: Creation of an arbitrary waveform with a very coarse granularity.

The value that we use for granularity must be an integer. It will be manipulated together with our desired output frequency (which we will tackle next) to generate the FSMP instruction as described above. The more points there are in the waveform, the longer it will take to download as every point in the waveform must be downloaded. The number of points that can be transmitted to the DS345 is very large indeed, but since the maximum number of lines in Excel, 65536, exceeds the maximum number of points, 16300, Excel should not be a limiting factor in this regard. The approach to this model will be to create a table for all the points. In an attempt to help with the comprehension of the model, I am going to be working with relatively small numbers: 100, 500 and 1000, although the user can theoretically enter any number up to 16300. At higher frequencies, the DS345 also places some restrictions on the number of points as we shall see.

The following code is added to the Initiate procedure so that it is run once when the worksheet is opened. It creates the drop-down text that is used in the Granularity control. Keep in mind that the control works with text values that we will have to convert to values later.

```
Sheet1.Granularity.Clear
'numbers are added to existing which is
'especially a problem duuring developemnt
'as the items are added over and over again
With Sheet1.Granularity
  .AddItem "100"
  .AddItem "500"
  .AddItem "1000"
End With
'and inititalise the selection
Sheet1.Granularity.Text = "100"
```

It is also probably prudent to parse the output from the Granularity control, so we need to add the following to the Granularity_Change event. You can get to this event code from the VBA Explorer by double-clicking on the Sheet1 folder and then finding the object in the drop-down menus in the Events Box above the VBA Editor window. Alternatively, within Excel set the Controls Toolbox to the Design Mode (the Set Square icon in the top left-hand of the toolbox), right-click on the control and select **View Code**.

```
Private Sub Granularity_Change()
  If IsNumeric(Granularity.Text) = False Or Right(Granularity.Text, 1) = "." Then
  'this is monitored character by character entry
  'exclude non numeric characters and decimal point
  'ignore the last entry if non-integer entry
    If Len(Granularity.Text) = 0 Then
      Granularity.Text = ""
    Else
      Granularity.Text = Left(Granularity.Text, Len(Granularity.Text) - 1)
    End If
  End If
End Sub
```

In Parenthesis: *String Functions*

It is possible to analyze and manipulate text strings by using the LEFT, RIGHT and MID functions.

Left (String, NumberOfCharacters) will look at the specified number of characters on the left of the string. Similarly, *Right (String, NumberOfCharacters)* will look at the specified number of characters on the right of the string. It is possible to access the mid portion of the string using the MID function. Its format is:

Mid (String, Start, Length). The start is where the extraction will begin, and the length is the number of characters that will be extracted. If length is omitted, then all the characters to the end of the string are returned.

It is possible to find out how long the string is using the *Len* function.

Combining of strings can be achieved with the concatenation action, which is simply expressed using the ampersand "&" character to link the strings. The plus symbol "+" fulfils the same function. Within Excel, there is also the CONCATENATE function which also does the same thing.

The *InStr* function will locate a string within a second string based on its position from the beginning of the string. The *InStrRev* does the same, but measures the position from the end of the string.

StrComp will compare to see if strings are equal and can check for equality based on text and binary data. See the VBA help file for greater detail.

Adding VBA Controls: Frequency

The frequency we will be entering is obviously the frequency of the output waveform. As engineers, we are accustomed to express the frequency in Hz, kHz and MHz, so we should allow that approach here by means of two Combo Boxes.

Frequency can have any value as long as it is numeric. The approach is very similar to the Granularity using a text box and a combo box (called *Frequency*). The linked cell is at Workings!B2. We should note that the ListRows property should be set to greater than or equal to the number of entries in the drop-down list. Set this to 9. The Initiate procedure has the following setup added:

```
Sheet1.Frequency.Clear
With Sheet1.Frequency
    .AddItem "1"
    .AddItem "2"
    .AddItem "3"
    .AddItem "4"
    .AddItem "5"
    .AddItem "6"
    .AddItem "7"
```

```
    .AddItem "8"
    .AddItem "9"
End With
'and inititalise the selection
Sheet1.Frequency.Text = "1"
```

The Frequency_Change event, the occurrence that is triggered by a change in the Frequency Combo Box, is basically the same as the Granularity event except that it will not exclude a decimal point:

```
Private Sub Frequency_Change()
    If IsNumeric(Frequency.Text) = False Then
    'this is monitored character by character entry
    'exclude non numeric characters
    'ignore the last entry if non-integer entry
        If Len(Frequency.Text) = 0 Then
            Frequency.Text = ""
            Worksheets(2). Range("fsp").Value = 0

        Else
            Frequency.Text = Left(Frequency.Text, Len(Frequency.Text) - 1)
            Worksheets(2).Range("fsp").Value = Val(Frequency.Text)
        End If
    Else
        Worksheets(2).Range("fsp").Value = Val(Frequency.Text)
    End If
End Sub
```

As we will see in a while, there will be another method of changing the frequency so I have created a cell named FSP at Workings!C2. The frequency set is copied to this location whenever the Frequency_Change event occurs.

The Units Combo Box can only have specific entries and Style property on this control should be set to 2, since only the entries shown in the drop-down list can be selected. The MatchEntry property should be set to 1, and the output linked to Workings!B3. The code added to the Initiate procedure is:

```
Sheet1.Units.Clear
With Sheet1.Units
    .AddItem "Hz"
    .AddItem "KHz"
    .AddItem "MHz"
End With
'and inititalise the selection
Sheet1.Units.Text = "MHz"
```

There is no need to parse the input since it is not possible for the user to enter any data. On the Workings sheet in cell C3, we add the formula:

=IF(B3="Hz",C2,(IF(B3="KHz",C2*1000,C2*1000000)))

which will calculate the actual frequency desired based on the number entered and the units selected (using the value in the FSP cell). In order to warn the user if the desired frequency is beyond the 30 MHz maximum, cell Controls!C5 is merged with C6 and C7 and formatted for red text. It contains the formula:

=IF(Workings!C3<=30000000,"","Maximum Frequency: 30MHz")

so that the message is displayed whenever the maximum frequency is exceeded (see Figure 16-5).

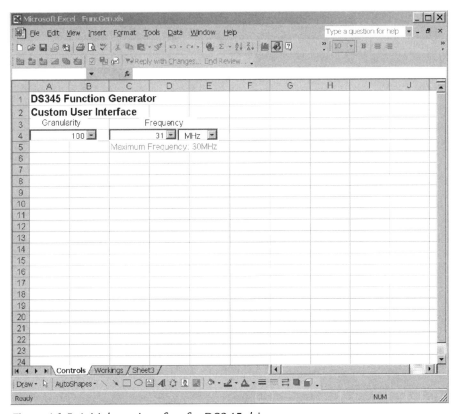

Figure 16-5: Initial user interface for DS345 driver.

Waveform Sampling Frequency

The granularity and the frequency must be combined to calculate the waveform sampling frequency needed by the FSMP command to the function generator. From the DS345 user manual definition, 1/FSMP is the time that the output value is held for before moving to the next value. For us that period is 1/(granularity*frequency), and so FSMP = granularity * frequency. However, there is a further constraint as part of the DS345 requirements. FSMP can only be an integer divisor of 40 MHz/N and cannot exceed 40 MHz, so it is obvious that at higher frequencies our granularity will drop down and our synthesized waveform will not be particularly smooth for anything other than a square wave. This means there will be a little give and take in order to arrive at a desired setting. First the user selects the granularity g_s and the frequency f_s. The initial stab at FSMP is:

$$FSMP_{init} = g_s * f_s$$

and this should be equal to 40MHz/N where N is an integer.

$$40MHz/N = g_s * f_s$$

Rearranging, we set N equal to the integer result (using the \ operator) of the division

$$N = 40MHz \setminus (g_s * f_s)$$

Obviously if N is 0, there is a problem and we will annunciate this. Using N, we now reverse the calculation process. I am going to assume that if the user has chosen a particular frequency, it should take precedence over the number of samples and remain constant. The new granularity g_n is given by:

$$g_n = 40MHz/(N * f_s)$$

The recalculation will be offered to the user to OK or Cancel. It is important to get the number correct because this determines the number of values of the waveform that will be calculated and downloaded. It must be an integer number, so the frequency will have to be modified as a consequence. The effective granularity g_e is derived from:

$$g_e = g_n \setminus 1$$

The effective frequency f_e is recalculated:

$$f_e = 40MHz/(g_e * N)$$

It seems to me that there may be a case for substituting this process with the Solve For feature, but I don't want to cloud this example with other issues as we may run into the integer issue that we discovered in Example 11. At any rate, the approach is consistent with the other parameters that we will add. So for better or worse, the procedure SetFSMP is implemented based on the above algorithm:

```
Sub SetFSMP()
    Dim TransmitBuffer As String
    Dim Response As Variant
    'calculating FSMPinit
    Range("FSMPinit") = Range("GS") * Range("FSactual")
    'and now N
    Range("N") = 40000000 \ Range("FSMPinit")
    If Range("N") = 0 Then
    'force the granualrity to be 1
        Range("N") = 1
    End If

    'create new granularity GN
    Range("GN") = 40000000 / (Range("N") * Range("FSactual"))
```

```
Range("GE") = Range("GN") \ 1
If Range("ge") = Val(Range("gs")) Then
'if no change force the update nevertheless
    Response = vbYes
Else
    'If Range("ge") = Range("gs") Then
    '   Response = vbYes
    'Else
        'if there is a change then notify
        Response = MsgBox("Granularity modified to " _
        + Str(Range("GE")) + Chr(13) + Chr(10) + _
        "Is this acceptable", vbYesNo, "New Granularity")
    'End If
End If
If Response = vbYes Then
    Range("FE") = 40000000 / (Range("GE") * Range("N"))
    Range("FSMPactual") = Range("ge") * Range("FE")
    Worksheets(1).txtGranular.Text = Range("ge")
    Worksheets(1).txtFrequency.Text = Round(Range("fe"), 2) & "Hz"

    Call SerialPortOpen
    TransmitBuffer = "FSMP" + Str(Worksheets(2).Range("c4")) + Chr(13)
    UserForm1.MSComm1.Output = TransmitBuffer
Else
    'place holder in case of need
End If
End Sub
```

You will notice towards the end of the code that I have added some Text boxes to show to the user the actual values used. These Text boxes have been added to the control interface and can be seen as the green text on a black background under the "Actual Frequency" and "Actual Granularity" headings in Figure 16-7.

I struggled to find a combination of events around the changes of the three combo boxes above, but because of the interrelationship, I couldn't find the right combination that would result in only one call of SetFSMP under all circumstances. I looked at the LostFocus, Key-Down and Click events, but in the end I solved this by creating a Command button. That allows the user to change all the values and then **Update** by clicking on the button. This button is also shown in Figure 16-7.

Bump Frequency

To add some versatility to the frequency change, I added a Scroll control that will allow the user to move the frequency by ± 50%. The properties of the control are shown in Figure 16-6. The linked cell is set to the cell Workings!B10. I have set the Maximum property to 50 (for 50%) and the Minimum property to –50. A SmallChange (by clicking on the

Up or Down arrows of the slider) will result in a change of 1 in the Value property of this control. A LargeChange will change by 5 (clicking in the space between the slider bar and the arrows). Of course moving the slider bar (a.k.a. *scroll thumb*) will result in a proportional change as well. Each of these will trigger a scrBumpFreq_Change event which will modify the value in FSP (overwriting the value placed there in the Frequency Change event), and then update the frequency output using the setFSMP procedure. Note that the maximum value is at the bottom of the slider, so I had to manipulate the operation to reverse this to have the maximum value at the top.

```
Private Sub scrBumpFreq_Change()
    Dim vTemp As Variant
    vTemp = Worksheets(2).Range("fs").Value * scrBumpFreq.Value / 100
    'multiply by -50 to 50 and work as percentage
    Worksheets(2).Range("fsp").Value = Worksheets(2).Range("fs").Value - vTemp
    'up = increase frequency in percentage
    Call SetFSMP
End Sub
```

Figure 16-6: Properties of the scrBumpFreq scroll bar.

314

When the frequency is entered via the Combo box, it should reset the scroll bar to the center position. This may automatically result in a SetFSMP call, so the command button click should be modified as follows:

```
Private Sub cmdUpdate_Click()
    If scrBumpFreq.Value = 0 Then
        Call SetFSMP
    Else
        scrBumpFreq.Value = 0
        'which will trigger setFSMP
    End If
End Sub
```

Figure 16-7 shows the results of our efforts so far.

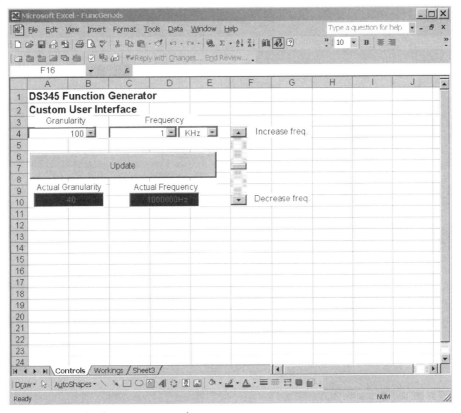

Figure 16-7: The frequency controls.

Generating Frequency Tables

The DS345 waveform is generated from a RAM table driving a 12-bit DAC converter. The input values to the DAC can range from −2047 to + 2047. The maximum range of the DAC is thereafter amplified to the desired amplitude by a separate amplifier that is calibrated so the waveform with a peak value of 2047 would have defined output amplitude. In other words, any waveform we create should aim to peak at 2047. If the DS345 is set to an amplitude of 2V, the number 2047 would correspond to this output and –2047 to –2V. Changing the amplitude to 5V will associate the 2047 value with this output amplitude.

Rather than download a waveform directly to the Function Generator, I decided to create an Excel table so that we could generate a standard waveform and then manipulate it either by manually editing the table or creating a second waveform and adding the two together. Indeed, using a Fourier transform would allow you to combine a number of signals to create any waveform.

In Parenthesis: *Fourier Analysis*

I have absolutely no experience in Fast Fourier Transforms, so I cannot give you an example of how to use FFT in Excel, but the feature is supported. Click on **Tools | Data Analysis | Fourier Analysis**. Here is a quote from the Excel 2002 help file:

"The Fourier Analysis tool solves problems in linear systems and analyzes periodic data by using the Fast Fourier Transform (FFT) method to transform data. This tool also supports inverse transformations, in which the inverse of transformed data returns the original data."

I inserted three more Combo box controls on the Controls worksheet. The first, called *comWaveform*, allows for three possible waveforms as an initialization. They are *Sine*, *Triangle*, and *Square*, and are set up in the Initiate procedure. Obviously, you can create many other waveforms if you want.

One of the idiosyncrasies of the DS345 is that the signal always oscillates positive and negative about the time axis. Even setting a TTL output on the instrument simply applies an offset of 2.5V and sets the amplitude of the wave (irrespective of the waveform) to 2.5V. Tweaking the amplitude results in the amplitude changing, but the offset remains at 2.5V, so the minimum of the signal moves away from zero. I have some applications where I want to vary the amplitude while keeping the minimum at 0V. The second combo box, comStyle is my solution to this. comStyle has two possible selections: *About Zero* and *From Zero*. The former is the standard output, and the latter is the new mode that I have just discussed. When we generate the waveform for the former, the numbers will range from –2047 to + 2047. In the latter case, the numbers only vary from 0 to 2047. I know it halves the resolution and in addition, the amplitude set on the DS345 front panel will also be out by a factor of two, but the minimum stays at 0. Providing an amplitude control on the Excel front panel will solve this problem.

The third combo box, *comScale* is added to allow us to scale the D/A output to less than 2047. The reason for this is if we are summing a number of signals, we don't want the result to be greater than 2047. I have limited the possible values to click selections from the Combo box only, simplifying the code and event interactions considerably.

The following listing shows the procedure for generation of the sine wave. The example on the CD-ROM, *FuncGen.xls*, has the code for the other waveforms as well. Figure 16-8 shows how far we've come.

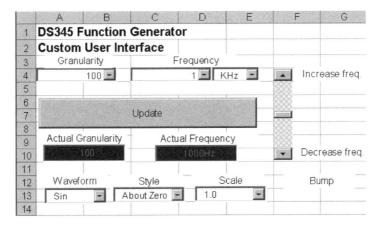

Figure 16-8: New additions to the User Interface.

The results of the waveform calculations are saved on the Workings worksheet starting at column F which represents the elapsed time in a single cycle, and column G which is the resulting waveform. The number of lines is determined by the Granularity. Column H allows you to add any data to distort a continuous waveform since columns G and H are summed for the result in column I. You should note that the formula for the summing of each column cell is added as part of the procedure so that there are only as many formulas as there are entries. The formula not only sums the cells in the two columns, it also caps the output to a maximum of 2047. Since the formula is inserted through software, the user can update the output cells in column I with impunity to create any waveform. Anytime the GenerateWave procedure is run, this information will be overwritten with the formula.

```
Sub GenerateWave()
    Dim iMin As Integer
    Dim iMax As Integer
    Dim il As Integer
    Dim varPeriod As Variant
    Dim strTmpStr As String

    'The range select that follows appears
    'to only work if the sheet is active
    'Clear ranges
    Worksheets("Workings").Activate
    Worksheets(2).Range("ChartTable").Select
    Selection.ClearContents

    'calculate period
```

```
    varPeriod = 1 / (Range("fe") * Range("ge"))

    If Range("Waveform") = "Sin" Then
       'if sin wave
       If Range("Style") = "From Zero" Then
          iMin = 1023 * Range("Scale")
          iMax = 1023 * Range("Scale")
          'offset adjustment
       Else
          iMin = 0
          iMax = 2047 * Range("Scale")
       End If
       For il = 0 To Range("ge") - 1
          'generate times in column F
          Worksheets(2).Range("a1").Cells(il + 1, 6) = il * varPeriod
          Worksheets(2).Range("a1").Cells(il + 1, 7) = Round _
             ((iMax * Sin(2 * 3.14159 * Range("fe") * _
             Worksheets(2).Range("a1").Cells(il + 1, 6))) + iMin)
       Next il

    Else
       If Range("Waveform") = "Triangle" Then
          'triangle wave
       Else
          'square wave

       End If
    End If
'insert formula into column I
    Range("I1").Select
    ActiveCell.FormulaR1C1 = "=IF(RC[-2]+RC[-1]>2047,2047,RC[-2]+RC[-1])"
    strTmpStr = "I1:I" & Mid(Str(Range("ge")), 2)
    Range(strTmpStr).Select
    Selection.FillDown
End Sub
```

In Parenthesis: *Fill*

In this example, the code for the formula fill was derived from recording a macro. In it I use a different method of copying data rather than the copy/paste technique used throughout the book. Why am I introducing this now? Because there are many things that I still have to learn about Excel and I just found this as an incidental part of a question in *PC Magazine*. One of the problems in systematically finding new features is that the cascading menus in Excel are contextual and you have to have the right combination setup to see a particular menu.

Anyway, back to how to do this. Enter a formula into the first cell of a row or column (or both) of the cell where the formula is going to be copied. In this example, the first formula will be entered into cell I1 and this formula will then be copied to I2 through to cell I100. Block-select the area where the formula will be copied (in this example, I1 to I100). Then click on the menu item **Edit | Fill | Down**, and it is executed. The last selection of **Down** will differ depending on the selection you have made to **Up**, **Right** or **Left**.

The waveform will be updated in the comWaveform, comStyle, comScale change event similar to the following:

```
Private Sub comScale_Change()
    Call GenerateWave
    Worksheets("Controls").Activate
End Sub
```

The call is followed by the reactivation of the Controls worksheet since the GenerateWave procedure sets the worksheet to the Workings sheet. The two statements are also added to the cmdUpdate button click and to the scrFreqBump change event.

Add a Chart

We saw in Example 14 how it is possible to create an open-ended chart, so that it dynamically accommodates different lengths of data. If we add a chart for the amplitude versus time for the single cycle, we could get an idea of what the waveform would look like before downloading it to the Function Generator.

On the Workings worksheet, select cells F1 to F100 and cells I1 to I100. Invoke the Charts Wizard, select the **XY (Scatter)** chart type and the **Scatter with data points connected by lines without markers** (the bottom right-hand option) sub type. Cosmetically modify the chart to your liking. I simply removed the **Show Legend** option in the **Legend** tab and I placed the chart on the Controls sheet. We should be looking at something like Figure 16-9.

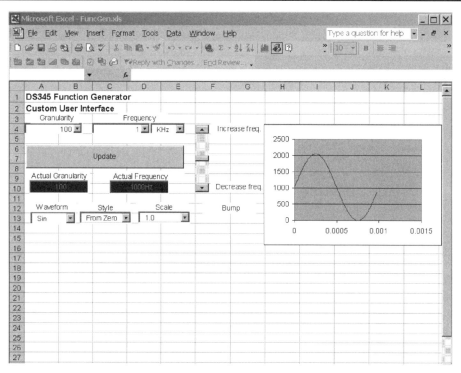

Figure 16-9: Control worksheet with chart derived from waveform data on Workings sheet.

Now we have to modify the chart to allow for input variation. Define a range by clicking on **Insert | Name | Define**. Add the name *time*, and define the range as:

=OFFSET(Workings!$F$1,0,0,COUNTA(Workings!$F:$F)-1)

as shown in Figure 16-10. For details on these functions used, refer to Example 14.

Figure 16-10:
Naming a
range.

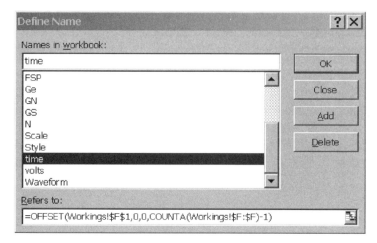

In a similar manner, add a second range named *volts*, and enter the range as:

=OFFSET(Workings!$I$1,0,0,COUNTA(Workings!$I:$I)-1)

Both of these ranges dynamically change their size by counting the number of entries within the range.

Click on the actual curve of the chart, and in the main formula bar modify the existing formula to read:

=SERIES(,FuncGen.xls!time,FuncGen.xls!volts,1)

This is shown in Figure 16-11.

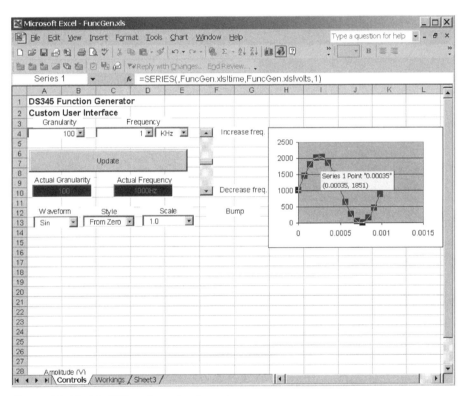

Figure 16-11: Modifying the series formula.

Try changing the waveform type, frequency or any of the other controls to see the chart dynamically update. Remember, some of the controls only update when the Update button is clicked.

As I have said before, I have designed this application so that it would be possible to go into the chart, modify the values and see the changes reflected as they are made. With the chart on the Controls sheet, this would require toggling between sheets. The same is true for changes in the settings of the controls. Although it is possible to place a chart in a number of places including other documents, I have not found a way to place it on a user form so that it will "hover" above the application and be visible all the time. However, it is simple enough to copy the chart from the Controls page and paste it in the Workings worksheet. Figure 16-12 shows a portion of the Workings sheet.

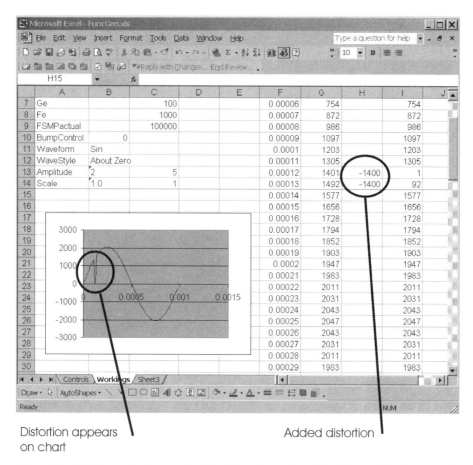

Distortion appears
on chart

Added distortion

Figure 16-12: Dynamic chart on the Workings sheet. Note that I have added two entries in cells H13 and H14 to distort the waveform and to see the immediate effect on the chart.

Download Waveform

In order to download, the commands must be given as text and then the data must be downloaded as binary. The data for each point in the chart is an integer number that is split into 2 bytes. In addition, a checksum is created by adding all the words and creating an integer that is transmitted as 2 bytes to complete the transmission. This is fully documented in the DS345 user manual. As you can see from the "In Parenthesis: VBA and Bit Manipulation" box, there are some software hardships in making these conversions.

In Parenthesis: *VBA and Bit Manipulation*

VBA has many advantages and great versatility except when it comes to bit manipulation. There is not much information around so what I show here, I arrived at by trial and error. It appears that the Byte variable type works as expected, that is, a division by 2 will shift the bits to the right, and a multiplication by 2 will shift the bits to the left. The problems start as VBA becomes overprotective. If the shift results in a number greater than 255, an overflow error is generated.

When you move to integer arithmetic, VBA will detect a problem when you try to set the most significant bit because (I am guessing) it expects that this is the negative sign and so you may have to end up generating complementary numbers. I reverted to bytes since VBA will allow you to set bit 7 of a byte type without generating a run-time error.

The final problem that I detected (and I cannot explain why my solution works) is that when working with more significant nibbles, dividing by 256 would return a number that was consistently out by one. If a 13 (0xd) was expected, the result would be 14 (0xe). Following are the steps that appear to generate the correct result:

1. AND the number with a mask of only the bits that we are interested in. For instance, for bits 8 to 15 use NUMBER And &HFF00,

2. Then perform the division: NUMBER/256,

3. And then look only at the relevant resultant bits: NUMBER And &HFF,

4. "cast" the number by storing in a byte variable.

VBA has a number of Type Conversion Functions. Although none of them appear to be of help in this instance, you can review them in the VBA Help. Just search for "Type Conversion".

The following code will generate the bytes and checksum and then download the waveform to the Function Generator. It does not transmit the amplitude.

```
Sub Download()
    'send LDWF?0,and number
    'wait for return
    Dim TransmitBuffer As String
    Dim varReturnedData As Variant
    Dim bytWaveData() As Byte
    Dim varWavedata As Variant
    Dim lngCheckSum As Long
    Dim il As Integer
    Dim intSpare As Integer
    Dim intSpare1 As Integer
    Dim varIntermediate As Variant
    Dim cc0 As Byte
```

```
ReDim bytWaveData((Range("ge") * 2) + 2)
'2 bytes for every value and checksum

lngCheckSum = 0

For il = 0 To (Range("Ge") - 1)
   varIntermediate = Worksheets(2).Range("a1").Cells(il + 1, 9)
   If varIntermediate < 0 Then
   'catering for the negative arithemtic
   'and forcing into an integer
      varIntermediate = varIntermediate * -1
      intSpare = varIntermediate
      intSpare = intSpare * -1
   Else
   'force into an integer
      intSpare = varIntermediate
   End If

   bytWaveData(il * 2) = intSpare And 255
   'least significant byte first
   intSpare1 = intSpare And &HFF00
   intSpare1 = intSpare1 / 256
   bytWaveData((il * 2) + 1) = intSpare1 And 255

   lngCheckSum = lngCheckSum + intSpare
   'maintaining checksum
Next il
'changing type
bytWaveData((Range("ge") * 2)) = lngCheckSum And 255
bytWaveData((Range("ge") * 2) + 1) = ((lngCheckSum And &HFF00) / 256) And 255

'initially we work in text mode
UserForm1.MSComm1.InputMode = comInputModeText
Call SerialPortOpen
TransmitBuffer = "LDWF?0," + Str(Range("ge")) + Chr(13)
'DS345 will ignore the space in str conversion
'TransmitBuffer = "LDWF?0," + Str(5) + Chr(13)
UserForm1.MSComm1.Output = TransmitBuffer
   'set to one character when read by Input

'rather than use the interrupt for a singel byte back
UserForm1.MSComm1.InputLen = 1
Do
   DoEvents
Loop Until UserForm1.MSComm1.InBufferCount > 1
varReturnedData = UserForm1.MSComm1.Input
```

```
If varReturnedData = "1" Then
   'it passes
Else
   'erroneous return,
   'but I am not sure what to do with it
End If

'now for the data
UserForm1.MSComm1.InputMode = comInputModeBinary
varWavedata = bytWaveData()
'placing array in variant
UserForm1.MSComm1.Output = varWavedata

'and then set the mode
UserForm1.MSComm1.InputMode = comInputModeText

TransmitBuffer = "FUNC5" + Chr(13)
UserForm1.MSComm1.Output = TransmitBuffer
End Sub
```

I decided that this procedure should only be run from a specific user button command as the detection of a user-induced change could be quite complex. Since the user could be working on either sheet, I placed Command buttons on both sheets and the click event of either simply calls the Download procedure. Figure 16-13 shows the resulting output on an oscilloscope.

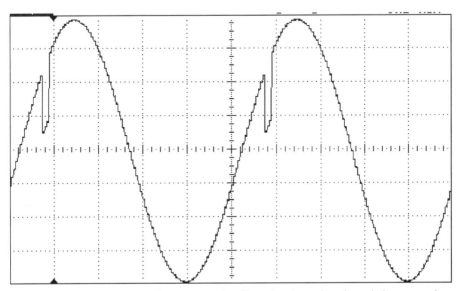

Figure 16-13: Output waveform. Note the distortion introduced, and the stepped output due to the fairly coarse granularity.

Setting the Amplitude

Rather than using the approach of the DS345 Function Generator of defining the output as amplitude, I chose to work with volts peak-to-peak since this means we don't have to consider the moving midpoint when we are working with a signal with the minimum fixed at 0V.

I added two Text boxes and a Scroll Bar as in Figure 16-14. Any change in the scroll bar value is transmitted to the DS345. The scroll bar is actually set up in percent (minimum of 0 and maximum of 100) because the output will change from 20 Vpp maximum for a signal about the time axis, and 10 Vpp maximum for a signal from the time axis. The resultant voltage is also displayed in one of the text boxes (the other acts as a heading). The code for the scroll bar change looks like this:

```
Private Sub scrVpp_Change()
    Dim varVpp As Variant
    varVpp = (scrVpp.Max - scrVpp.Value) / 100
    'as a %
    If Worksheets(2).Range("Style") = "From Zero" Then
        Worksheets(2).Range("vpp").Value = varVpp * 10
    Else
        Worksheets(2).Range("vpp").Value = varVpp * 20
    End If

    txtVpp.Value = Str(Worksheets(2).Range("vpp").Value) & " V"
    Call SetAmplitude
End Sub
```

and the procedure call to send the amplitude is found in Module1:

```
Sub SetAmplitude()
    Dim varTemp As Variant
    If Range("Style") = "From Zero" Then
        varTemp = Range("Vpp").Value
    Else
        varTemp = Range("Vpp").Value / 2
    End If

    Call SerialPortOpen
    TransmitBuffer = "AMPL" & Str(varTemp) + "VP" + Chr(13)
    UserForm1.MSComm1.Output = TransmitBuffer

End Sub
```

SetAmplitude is also added in the two Update buttons associated procedures.

Skew

Since some signal generators have a skew feature for sine waves, I am sure it is possible to express the signal mathematically, but I have been unable to find this relationship. It can be done by analyzing periods similar to the triangular wave approach that you will soon see. Since I don't want to bloat this example and still get a sub-optimal wave, I decided to disable the skew feature for sine waves and only use it on a square wave or a triangular wave. When a sine wave is selected, the skew controls are made invisible. For the other two options, the skew percentage control is enabled under the control of a Toggle button.

In order to implement the Skew control, the Waveform_change event is modified to the following:

```
Private Sub comWaveform_Change()
    Select Case Worksheets(2).Range("waveform")
        Case "Sin"
            scrSkew.Visible = False
            togSkew.Visible = False
            txtSkew.Visible = False
            togSkew.Value = False
            scrSkew.Value = 50
        Case Else
            scrSkew.Visible = False
            togSkew.Visible = True
            txtSkew.Visible = False
            togSkew.Value = False
            scrSkew.Value = 50

    End Select
    Call GenerateWave
    Worksheets("Controls").Activate
End Sub
The event triggered by clicking the Skew toggle button is:
Private Sub togSkew_Click()
    If togSkew.Value = True Then
        scrSkew.Visible = True
        txtSkew.Visible = True
        scrSkew.Value = 50
    Else
        scrSkew.Visible = False
        txtSkew.Visible = False
    End If
End Sub
```

Let us consider how this is implemented for a triangular wave. The code is:

```
Sub SetupTriangle(varPer As Variant)
    Dim iMin As Integer
    Dim iMax As Integer
    Dim il1 As Integer
    Dim varSlope1 As Variant
    Dim varSlope2 As Variant
    Dim varConst1 As Variant
    Dim varConst2 As Variant
    Dim varTimePeak As Variant
        If Range("Style") = "From Zero" Then
            iMin = 1023 * Range("Scale")
            iMax = 1023 * Range("Scale")
            'offset adjustment
        Else
            iMin = 0
            iMax = 2047 * Range("Scale")
        End If
        varTimePeak = (Range("ge") - 1) * varPer * Sheet1.scrSkew.Value / 100
        'finding where the peak period is
        'setting constants for the line
        'so they don't have to be recalculated for every il
        varSlope1 = 2 * iMax / varTimePeak
        varSlope2 = -2 * iMax / (((Range("ge") - 1) * varPer) - varTimePeak)
        varConst1 = -iMax
        varConst2 = iMax - varSlope2 * varTimePeak
        'using the waveform y=Slope*x + Constant
        For il1 = 0 To Range("ge") - 1
            'generate times in column F
            Worksheets(2).Range("a1").Cells(il1 + 1, 6) = il1 * varPer
            'now looking for which part of the waveform we are in based
            'on the skew setting
            If Sheet1.scrSkew.Value >= ((il1 / (Range("ge") - 1)) * 100) Then
                'on the left of the skew point
                Worksheets(2).Range("a1").Cells(il1 + 1, 7) = _
                Round((varSlope1 * Worksheets(2).Range("a1").Cells(il1 + 1, 6)) _
                + varConst1 + iMin)
            Else
                'on the right of the skew point
                Worksheets(2).Range("a1").Cells(il1 + 1, 7) = _
                Round((varSlope2 * (Worksheets(2).Range("a1").Cells(il1 + 1, 6)) - varTimePeak) _
                + varConst2 + iMin)
            End If
        Next il1
End Sub
```

Before I describe this, I should mention that there appears to be a shortcoming in the debugging capabilities of VBA. Stepping through some of these statements may cause VBA to report phantom incorrect references to a sheet (without actually specifying which sheet). If you run through the statements using breakpoints instead of stepping, the statement is executed correctly.

The first part of the procedure determines the limits of the waveform in the y dimension determined by the scaling and whether the signal oscillates about the time axis, or has its minimum set to 0.

The triangular waveform is divided into two lines of the form $y = mx + c$ where m is the slope and c is a constant. The next stage calculates the slope and the constant based on the maximum and minimum of the waveform and the position of the apex of the triangle defined by the skew setting. If the skew Toggle button is off, the skew setting is 50%.

The final step evaluates which line is to be used for a particular time, and calculates the associated value placing the time in column F and the result of the calculation in column G on the Workings sheet.

This procedure is called from within the GenerateWave procedure, and so on the return to that procedure the summing formula is added in column I and it is possible to add individual data points in column H to distort the waveform.

Figure 16-14 shows the controls and the resultant skewed triangular wave.

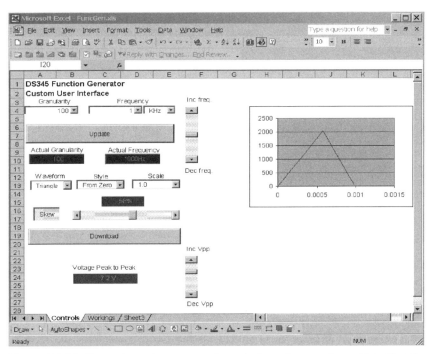

Figure 16-14: Controls used to setup a skewed triangular waveform. Note that the gridlines have been turned off (in the Options menu) to improve the appearance.

The code for the square wave is trivial compared to the triangle wave and can be found in the SetupSquare procedure.

Average Voltage, RMS Voltage

As we have seen in Example 10, it is possible to use Excel to evaluate definite integrals by calculating and summing the area of consecutive trapeziums created by the curve. The average value of a waveform is given by:

$$V_{avg} = \frac{1}{T} \int_{t0}^{t0+T} v(t)\,dt$$

The RMS value is calculated from:

$$V_{RMS} = \sqrt{\frac{1}{T} \int_{t0}^{t0+T} v(t)^2\,dt}$$

The calculation of the area of each trapezium is applied to the formulas added during the waveform generation at the end of the GenerateWave procedure:

```
Range("J2").Select
ActiveCell.FormulaR1C1 = _
    "=((ABS(RC[-1])+ABS(R[-1]C[-1]))/2)*(RC[-4]-R[-1]C[-4])"
'now to create areas using trapezium method for rms power
'add two square of ordinals and average them and multiply
'by the timw
Range("K2").Select
ActiveCell.FormulaR1C1 = "=((RC[-2]^2+R[-1]C[-2]^2)/2)*(RC[-5]-R[-1]C[-5])"
strTmpStr = "J2:K" & Mid(Str(Range("ge")), 2)
Range(strTmpStr).Select
Selection.FillDown
```

The average value calculated is the value seen on a non-RMS Digital Voltmeter when measuring AC volts since the mathematical average of a sine wave is zero. The area of each trapezium is stored in column J, and the area of each trapezium of the squared value is stored in column K in the above listing. To find the definite integral, we need to sum all of these areas. In order to generalize the calculation, we need to name the columns with the open-ended approach used for the chart. We do this by creating two named regions using **Insert | Name | Define,** and create a region called *AveVolt* and enter the formula:

=OFFSET(Workings!$J$2,0,0,COUNTA(Workings!$J:$J)-1)

in the **Refers to** bar. Create a second region called *RMSVolt* with the following formula:

=OFFSET(Workings!$K$2,0,0,COUNTA(Workings!$K:$K)-1)

The reciprocal of the period is the frequency, and so we can add the formula for the average value in cell B17:

=FSactual*SUM(AveVolt)

For the normal sine wave about the time axis with an amplitude of 2047, this returns a value of approximately 1300. This is a factor of 1300/2047=0.635, which is very close to the actual value of 0.636. This factor is stored in cell C17 (named *AveFactor*). Note that there are two sources of error besides that quantization error. First, there is one reading missing, the last period of the waveform that would return the signal to its original value. Second, where the wave crosses the time axis (goes positive to negative and vice versa) there is a potential for further inaccuracy.

Similarly, we enter the following formula in cell B18 to calculate the RMS value:

=(FSactual*SUM(RMSVolt))^0.5

The calculated value for the sine wave is approximately 1447, which is a factor of 0.707 which is exactly the RMS value to the first approximation despite the inaccuracies discussed above. This factor is stored in cell C18 (named *RMSfactor*).

The Crest Factor (CF) is defined as the peak voltage divided by the RMS value. Since the conversion to voltage would be applied to numerator and denominator and would cancel out, we can calculate the CF in cell B19 using the following formula:

=MAX(OutSig)/RMS

where RMS is the name for cell B18, and OutSig is another named region define by:

=OFFSET(Workings!$I$1,0,0,COUNTA(Workings!$I:$I)-1)

For the sine wave example, the calculated value is 1.414 which is exactly right.

It would be nice to show the average voltage, RMS voltage and Crest Factor on the Control sheet. It would be pretty easy to process the numbers in a cell on the sheet, but it would not be consistent with the other displays. There are three Text boxes added, but the question is what event will update them. The best event, I thought was the calculate event for Sheet2. After some debugging, the code evolved into:

```
Private Sub Worksheet_Calculate()
    'when this is recalculated, update
    'the average, RMS and CF on Controls sheet
    Dim varTemp As Variant
    Dim varFactor As Variant

    If IsNumeric(Range("avefactor")) = True _
    And IsNumeric(Range("RMSfactor")) = True _
    And IsNumeric(Range("CF")) = True Then
        If Range("Style") = "From Zero" Then
        'determining divisor
            varFactor = 1
        Else
            varFactor = 2
        End If
```

varTemp = Round(Range("Avefactor") * Range("scale") * Range("vpp") / varFactor, 3)
Sheet1.txtAvg.Value = Str(varTemp)
varTemp = Round(Range("RMSfactor") * Range("scale") * Range("vpp") / varFactor, 3)
Sheet1.txtRMS.Value = Str(varTemp)
varTemp = Round(Range("CF"), 3)
Sheet1.txtCF.Value = Str(varTemp)

End If
End Sub

The calculate event may happen several times as a result of some action. The problem is that during the intermediate stages, the calculated factors sometimes return faults that are cleared when the process is complete. In order to prevent VBA from detecting an error we need to check that the accessed cells do indeed contain a number (and hence not an error message) before allowing the text boxes to be updated.

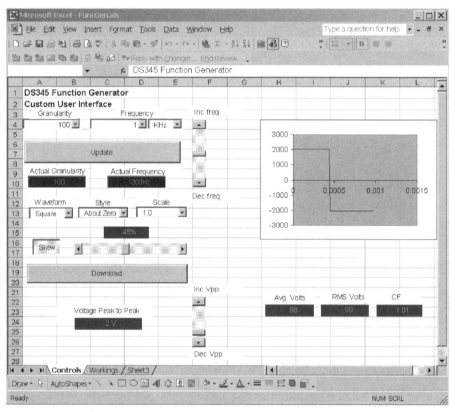

Figure 16-15: The completed user interface.

Ah, the sweet science of ergonomics! As you can see, my layout in Figure 16-15 could benefit from some help, but there is no denying how versatile the combination of Excel and the DS345 Function Generator can be in generating custom waveforms.

APPENDIX A

VBA and Excel

Since it is a standard across all Microsoft Office applications (and now other applications), VBA is obviously consistent in all those applications except in the nitty-gritty details of how it interacts with the specific objects within the application. It was never my intention to provide a detailed VBA description, since this book was supposed to show by "example." However, I feel there should be some reference point where the basics are collected in one place. This is that place.

Where Do Macros, Procedures and Functions Get Stored?

Macros, procedures, functions, forms and so forth are stored within a workbook, although not necessarily the workbook you are working on. There are several methods to access code depending on your application. First, let's consider where this code is stored in the workbook.

Figure A-1 shows some aspects of the VBA (Project) Explorer which is part of the Visual Basic Environment. Within each workbook are a number of objects as can be seen in the hierarchical structure. Double-clicking on the object will result in the information pertaining to that object being brought into "focus." If the object is a form, the form is displayed. If it is some code, then the code is displayed.

The Module view buttons will change how code is displayed in the code window, showing a procedure at a time or as one long document. An object may have different elements and each can be selected in the drop-down Object box. The procedure within the object element is selected from the drop-down Procedure box.

Any code is stored in a module or on a sheet. Typically, the code associated with events on the sheet is stored there, but it is possible to enter any procedure there as well. Macros are stored in modules. Recording a macro automatically opens a module and records it in that module. Depending on what modules are open, their names, and what is contained in the module, VBA may or may not open a new module. I have not found a rule as yet.

You can change the name of a module, and this is often advisable when there are going to be two or more workbooks open. Click on the module that you want to rename. If a

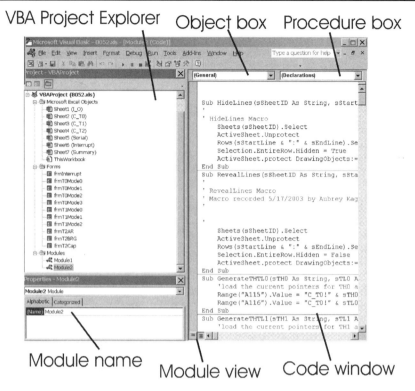

Figure A-1: VBA Explorer.

"Properties–ModuleX" window does not appear, follow **View | Properties Window**, or press <**F4**>. Then change the name in the (name) field.

Procedures may be grouped logically in multiple modules in preparation for using the module in other applications. Moving the procedures around can be achieved with standard cut and paste, or drag and drop techniques.

Opening a second (or more) workbook results in the new workbook extending the tree within the VBA Explorer, so moving modules across workbooks is pretty easy.

Using a Macro

If a macro is only applicable to a particular workbook, the macro need only reside in that workbook so there is no further consideration required.

Personal Macro Workbook

If there are certain macros that you always want access to, saving to the Personal Macro Workbook saves this to a hidden file called *Personal.xls*. This option is found in the Record macro dialog box as shown in Figure A-2. The file is created the first time you write to it and is saved in the XLStart folder which will be located deep in the hierarchy of the hard drive. Mine is located at C:\Program Files\Microsoft Office\Office10\XLStart. If you want to edit or delete a macro from the Personal Macro Workbook, you first need to unhide the workbook using the **Windows | Unhide** (in Excel) sequence. The sequence is reversed to hide it again, since it will not become invisible automatically.

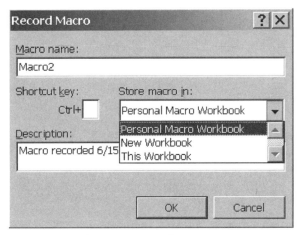

Figure A-2: Selecting storage of macro.

Open Workbooks at Startup

Any workbook in the XLStart folder is automatically opened when Excel is started, so you can create a number of workbooks that you want to open on initialization. The Personal Macro Workbook is only different in that it is hidden automatically.

If your Excel application is a shared application, the XLStart folder is a shared folder as well. It is possible to add a second startup folder; first, create the folder on your drive and then in Excel click on the sequence **Tools | Options** and go to the **General** tab (see Figure A-3). Enter the directory in the box titled **At startup, open all files in:**.

Accessing a Function Across Workbook Boundaries

As can be seen in some of the applications in this book, it is possible to access a function across workbook boundaries. The first method needs to be done once for a workbook and is accessed from the VBA application. Go through the following menu sequence **Tools | References | Browse** and then find the target workbook that contains the functions that you want to access. If you use this approach, it is recommended that the project and the module in the target workbook be named to some unique name. If not, an error message will be generated. Even if you use this method, the upshot will be that the user will be prompted to open the target workbook if not already open.

The second method requires that the target workbook be open. This can be done manually or through code in VBA. Checking for open workbooks and opening them is discussed a little later in this appendix.

For both approaches, in order to access the target functions, click on the cell where you want the function to reside, select the menu sequence **Insert** | **Function** | **User Defined** and select the function required from the presented menu.

I am not entirely sure why the first method exists, since in either case, any time you reopen the principal application it will insist on the support application being open. I investigate this in Example 10.

If a macro in another workbook is linked to a custom toolbar, when the workbook with the toolbar is opened, the second workbook (with the macro included) is also opened.

Template

Sometimes we may want the template of a workbook, which may include all the necessary code for the application. Once the template is opened, all association with the new workbook is terminated so that any further changes in the new workbook do not appear in the template. To create a template, save the worksheet in the templates folder which is normally at:

 C:\Documents and Settings\user_name\ApplicationData\Microsoft\Templates.

By clicking on **File** | **Save as** | **Save as Type** | **Template** and then entering a name (if different from the actual file name) and clicking on **Save**. Every time a new workbook is started, the user will be given the option of using this file as a template.

Figure A-3: Adding a startup folder.

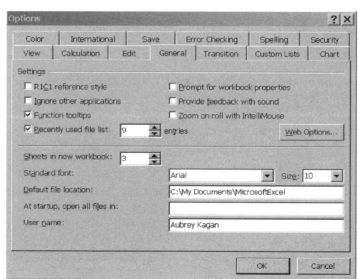

If you want to modify the template that is used for the blank new workbook, it is possible to change all or part by creating one or more of the following templates: *book.xlt*, *sheet.xlt*, *chart.xlt*, *dialog.xlt*, or *module.xlt* and saving it in the XLStart folder or the alternate as described above.

Add-Ins

With some of the functions that you create, it would be ideal to have them available as a library to use as and when necessary, exactly as with the libraries provided for Excel. This means that the library has to be loaded first (**Tools | Add-Ins**), and then the procedures/functions are available for all installation in any workbooks at any stage The advantage is that this method adds a level of security in that the code is compiled and can be made invisible and inaccessible to the user.

Step 1. Once the workbook with the desired functions is complete, it must be compiled. In the Visual Basic Environment, click on **Debug | Compile VBA Project**.

Step 2. Right-click on the workbook in the Visual Base Explorer. In the dialog window (Figure A-4) add a project name. This is the name that will appear in the menu when it is added. Click on the **Protection** tab and enter the required information to prevent a user from viewing it. Incidentally, a workbook can also be protected in this way. This is not the same as the **Tools | Protection** option in the workbook.

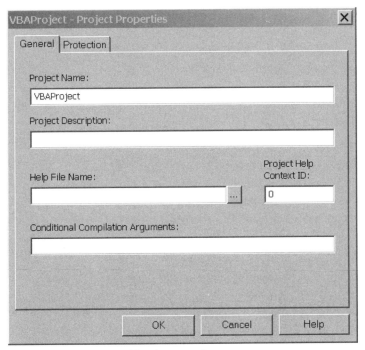

Figure A-4: Setting workbook properties.

Step 3. Using the menu sequence **File | Save As** and selecting the **Microsoft Excel Add-In (*.xla)** option in the **Save as type** options, save this to anywhere you like.

Provided you have the password to open the add-in, it can be debugged without having to use the original .xls file.

It should be noted that if an add-in has been added under VBA control as part of a procedure, closing Excel will not close the add-in. You will need to add a procedure to do this in the "Workbook_BeforeClose" event.

Automatic Startup

If a procedure is named *Auto_Open*, then it is executed when the workbook is opened. This is not the only way of doing this. There is an event called *Workbook_Activate*, which obviously occurs in that circumstance. The startup procedure could be located here.

There are instances where you may want the computer on power up to go straight into Excel and execute a particular process. You could achieve this by placing a shortcut to Excel in the Windows startup folder, placing the desired file in the XLstart folder and then setting up the procedure to run in the workbook_activate.

If you don't want the "Workbook_Activate" macro to run, hold the <**Shift**> button down while the workbook is starting up.

Private

Use of the word "Private" in the procedure declaration has other VBA implication as to the visibility of variables and so forth, but it also means that the procedure cannot be seen from the **Tools | Macro | Macros**. This means that an inexperienced user will not be able to run them by accident. Some procedures are only called from other procedures. It is probably prudent to classify them as private so that they cannot be run before the calling routine has set up anything that is needed.

Running a Macro

Excel is an event-driven application. Any click, change of value, selection of a cell, and so forth triggers an event. Once you have located that event within the VBA environment, any procedure or function can be executed in that event.

In the macro dialog, it is possible to associate a macro with a "hotkey" combination as well (see Figure A-2). Take care not to overwrite an existing key combination like <**Ctrl**> + <**V**>.

Certain objects like buttons and menus have dedicated menu entries for adding a macro when you right-click on them.

Finally, a macro can also be run when called from VBA (and from the debugger of course).

Finding the Stop Recording Macro Toolbar

The stop recording macro toolbar initially pops up when you record a macro. It is all too easy to click on the **X** at the top of the window and it disappears forever. In order to make it reappear, go through the menu sequence **Tools | Customize**, select the **Toolbars** tab and make sure the **Stop Recording** selection is checked.

Actions Recorded During Macro Record

Once you have recorded and analyzed a few macros, you will realize that the feature generates a generic approach, and there may be many statements that appear in the macro that do not change anything. You can easily delete these.

Names

In Excel, it is possible to name cells, a range of cells, an object and even a row or a column. Using the same technique, it is possible to create constants and name a formula.

Naming Objects

To name an object (for instance, a chart), click on the object and then in the name box enter the name. No other technique (such as **Insert | Name | Define**) will work in this instance.

Naming Constants

If you want to use a constant throughout the workbook without using a cell to hold the value, you can name a constant. For instance, if you wanted to use pF to indicate 10^{-9}, you could do the following:

Insert | Name | Define, which will bring up the Define Name dialog, which is filled in as per Figure A-5. This is similar to the definition of any constant in software and it can be used anywhere. To enter 220 pF as a value, you could enter a formula $=220*pF$, and the number is immediately calculated.

Figure A-5: Naming a constant.

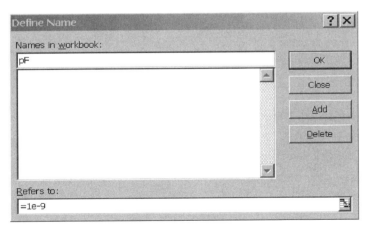

Naming Formulas

If you have a formula that is repeated several times throughout the workbook, always performing the same function on some cells (either absolute or relative), it is possible to name this formula. This is especially convenient if the formula is likely to change, since it only needs to be changed in one place and it is updated throughout the workbook. As an example, let's assume that we want to calculate the power dissipated in a resistor. Initially, we are considering DC and we have the current in cell H4, the resistance in cell I4, and we want the power shown in cell J4. We also assume that there are several resistors in the project and that they are all arranged in the same pattern in the cell. As we all know, the power through the resistor is I^2R. We click on cell J4 and go through the name definition as above. As the name enter, "*PowerDissipation*" and in the **Refers to:** box enter "$= H4\wedge2*I4$".

Now if you ever need to evaluate the power dissipation for AC current, all you have to do is edit the definition in the Define Name box, to account for the $\sqrt{2}$ factor.

Absolute and Relative Reference

Included in the Stop Recording macro toolbar is a button for absolute and relative reference of cells. It can be toggled on and off as necessary. Careful consideration should be paid as to what the macro is going to be used for. Let's say you wanted to move the cursor two cells to the right and use this as a generic approach (as we did in Example 8 by identifying the resistor color bands). In the absolute mode, the movement of the cursor simply results in code that makes a particular cell active:

Range("F7:F7").Select

This makes calling a generic procedure difficult. The use of the offset method allows movement of the cursor by rows and columns as in:

ActiveCell.Offset(m,n).Range("A1").select

When *m* and *n* are the row and column offsets and may be negative, allowing movement in any direction.

Normally, when using the point-and-click technique of showing Excel where you want the data to be fetched from or sent to, the resulting formula will be recorded in absolute format. Take care if you intend to copy this formula elsewhere. Excel is not always consistent in this approach.

Moving Cell Selection Relative to Current Cell

While we are here, it should be noted that the activation of a cell relative to the current selected cell can be done as above, or using the statement:

ActiveCell.Offset(rowOffset:=m, columnOffset:=n).Activate

When *m* and *n* are the row and column offsets and may be negative.

Cell Access

Many times it is not worth the effort to change the active cell just to access it. The active cell can be left untouched using the following techniques.

The cells may be accessed using the Range or Cells property. In most of the applications in this book, the cell references are a combination of the column letter on row number. This is called *A1 notation*. There is a setting in the Options menu (under the General tab) that allows an alternative notation of RnCn where any cell is denoted by a combination of the letter "R" (for Rows) and a number, and the letter "C" (for columns) and a number. The cells property is a variation of that although the RnCn option does not have to be selected to use the cells technique. I try to ignore the RnCn notation in this book, but occasionally the macro learn slips something in like the FormulaR1C1 property (see later).

The range property uses the following format:

Range{"Mn") where Mn can be a single cell in the standard letter/number format (A1 format), a range of cells formed by the top left-hand and bottom right-hand or a named range. An example might be:

Range("A3:B9").value=""

The cells approach deals with numeric values only so it is much easier to use in a loop statement. For instance:

for i=3 to 9

cells(i,7)="

end for

Sometimes, a sort of hybrid approach may be needed to define a range using the cell method, for example:

Range (Cells(5,6),Cells(10,12)).select

In the following example, a cell at the top left of a particular range is set to a specific value:

Worksheets(1).Range("F4:L10").Cells(1,1).value=9 will result in cell F4 having a value of 9.

If this cell were in another file you would enter:

Application.Workbooks("Wbook.xls").Worksheets(1).Range("F4:L:10"),Cells(1,1).value=9

As discussed above, you can also access a cell relative to another using the offset property. Changing the value of a cell three rows down and two across from a particular cell (without changing the selected cell), could be written as:

Selection.Offset(3, 2).Cells(3, 4) = 19

The current column can be found using the column property, and similarly, the current row from the row property. Changing the value of a cell relative to the current cell would become:

Selection.Offset(3, 2).Cells(ActiveCell.Row, ActiveCell.Column) = 19

It is possible to access the intersection of two ranges. The expression:

Range (F9:F15 D11:H11) accesses cell F11. The space between the ranges is the intersection operator.

In order to access a number of cells simultaneously, we need to use the union operator, which is the comma as in the following example: *Range ("A6, F7, X9")=72.*

To clear a group of cells, we could use the ClearContents method. To clear the entire worksheet:

ActiveSheet.Cells.ClearContents

Note that *cells* without the parenthesis accesses all the cells in the object. For a smaller range your could use:

ActiveSheet.Range("A1:F17").ClearContents

It is possible to persuade VBA to use the RnCn format by using the FormulaR1C1 property. Assume the active cell is G7.

ActiveCell.FormulaR1C1="R[-1]C[-2]" will get the value from cell E6.

Detecting Blank Cells

In this book we have used several techniques to detect if a cell is blank. We have looked for the length of a string (using the LEN function) to be zero. We have also checked that the contents were equal to "". Another technique is to use the function IsEmpty as in:

If IsEmpty(Range"A7") then …

Extended Range Selection

Range selection from the current cell to the last valid cell in the column can be achieved as follows:

Range (ActiveCell, ActiveCell.End(xlDown)).Select

The possible constants are *xlDown, xlUp, xlToLeft* and *xlToRight.*

Accessing Data off the Current Sheet

In the same workbook, preface the cell with the sheet name and an exclamation point (exclamation mark, in an alternative English dialect). For example, accessing cell Y3 on a sheet named "StandardCapacitors" would be:

> = StandardCapacitors!Y3 &"nF"

In a different workbook, this reference is prefixed by the workbook name in square brackets similar to this:

> =[Capacitors.xls] StandardCapacitors!Y3 &"nF"

If there are spaces in the filename, then single quotation marks must begin and end the full sheet access as in:

> ='D:\Excel Utilities\[Kemet Capacitors.xls] StandardCapacitors'!Y3 &"nF"

Provided the second workbook is open, it is possible to select the target using the standard point-and-click techniques. Keep in mind that changes in filenames can really confuse these settings. Updates while the files are open are normally reflected in the changes, but if there are accesses to a reference workbook from several other places, you have been warned! As a special caveat, keep in mind that using the **File | Save As** (especially as a backup procedure) changes the filename and the references to it.

Accessing a Procedure in a Different Module

If you have two procedures of the same name in different modules, you need to add a prefix using the module name to the procedure name, separated by a dot. Assuming Test1 exists in module Trial1 and module Trial2, you would need to enter:

> *Call Trial1.Test1*

Accessing a Procedure in a Different Workbook

This can be approached in two ways. It is possible to create the connection as described in "Accessing a function across workbook boundaries" above. Alternatively, you can do the following:

> *Call ProjectName.ModuleName.ProcedureName*

Or,

> *Application.Run "Filename.xls!Trial1"*

Note that if the filename includes spaces, it must be included in single quote marks. The file must be open to access the procedure.

Recalculate

Unless the feature is turned off, Excel updates cells as necessary when the precedents to the cell change. However, it is possible to force a recalculation in VBA as follows:

ActiveSheet.Calculate

Screen Update

Some procedures may take some time to execute. The execution time is lengthened by screen updates. These can be prevented by turning the updates off:

Application.ScreenUpdating=false

and then reenabling by setting the property to true.

Exit

It is possible to force early termination of the execution of a loop or a procedure using the Exit statement. For example:

Exit sub or

Exit for

Workbook Open, Activation

Certain actions can only be done if a workbook is open, or active or both. In order to detect if a workbook is open, we would need to do something like this:

```
Function bIsWorkbookOpen(sCheckFile As String) As Boolean
    Dim WB As Workbook
    bIsWorkbookOpen = False
    For Each WB In Application.Workbooks
        If WB.Name = sCheckFile Then
            bIsWorkbookOpen = True
            Exit For
        End If
End Function
```

This approach will determine any open workbooks including hidden workbooks, but does not include open add-ins. For an add-in, you have to examine whether it is open by checking for its name explicitly.

To activate an application requires a statement of the form:

WorkbookName.Activate

And to open a workbook, the form is:

Workbooks.Open WorkbookName

Where *WorkbookName* is the name of the workbook to be opened including the path, if necessary.

Multiple Actions on an Object: With Statement

There are times when a number of actions need to be done on a single selection. For instance, cell A23 could be set to the value 123.34, in Arial font and in bold with a font size of 12.

You could have:

```
ActiveSheet.Range("A23").Value=123.34
ActiveSheet.Range("A23").Font.Name="Arial"
ActiveSheet.Range("A23").Font.Size=12
ActiveSheet.Range("A23").Font.Bold=True
```

It would be better to write this as:

```
With ActiveSheet.Range("A23")
.Font.Name="Arial"
.Font.Size=12
.Font.Bold=True
End With
```

Using an Excel Function in VBA

There are some instances where it would be convenient to use a function that exists in Excel to save reinventing the wheel. It is possible to do this, provided that a function of the same name does not exist in VBA. For instance, the MOD function exists in both Excel and VBA.

In order to do this, we preface the complete Excel function with "Application.Worksheet-Function.". For example, to find the average of a range we would enter the line of code:

```
Variable=Application.WorksheetFunction.Avg(range("c15:c25"))
```

Where Variable obviously receives its value from the result of the function call.

Arrays

An array in Excel is purely a range of cells. It is possible to perform actions on an array by entering the formula once, and all the action is then performed on every element of the array. Those of us familiar with matrix operations (do they still teach that in the university?) will find this concept intuitive.

Array Formula

Let's consider the calculation of power in a circuit based on the current flowing through a component and the voltage across it. If Congress has not yet changed the laws of physics, power is the product of current and voltage. If we tabulate them as in Figure A-6, we can then create the array formula as follows: block the range C8 to G8 where the results will appear. Enter the "=" character in the formula bar, click and drag over the range C6 to G6 (noting the appearance of the range in the formula bar). Type the "*" character, and then click and drag the range C7 to G7. Now comes the secret to the array formula. Instead of hitting **<Enter>**, you must use the key combination **<Ctrl> + <Shift> + <Enter>**, and the answer appears in all the designated cells. In the formula bar, this calculation is indicated by the use of the curly brackets "{" and "}".

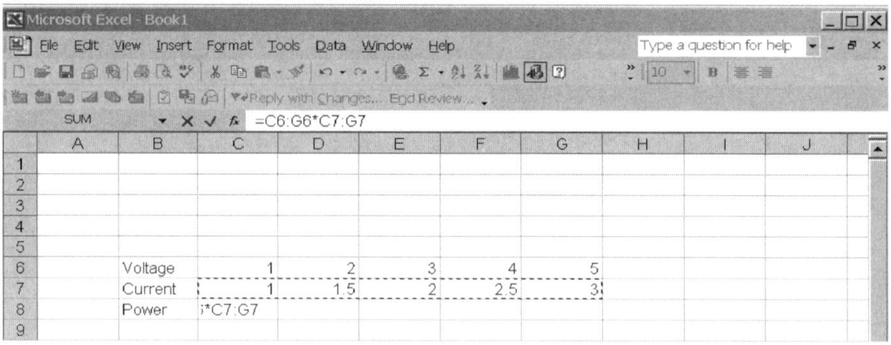

Figure A-6: Entering an array formula before.

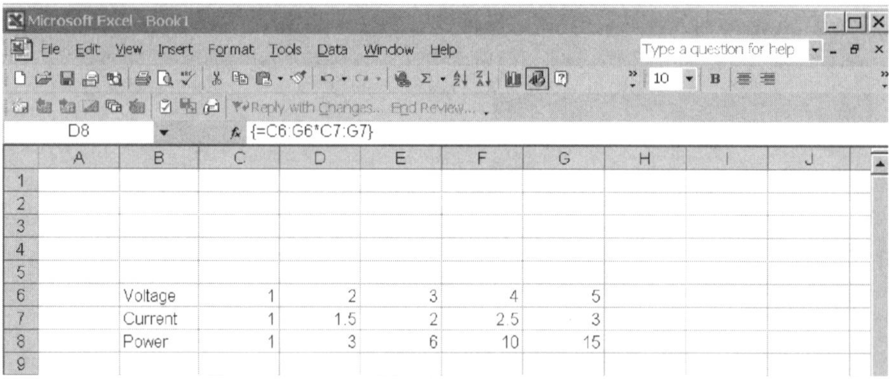

Figure A-7: Array formula, after <Ctrl> + <Shift> + <Enter>.

An interesting aspect of the array formula is that Excel can hold the array result in memory and perform and display a function applied to that array. Let's change this example to calculate the total power dissipation (Figure A-8). Click on the cell where the result will live, cell D10 in this case, and type into the formula bar (using click and drag techniques if you want):

=*sum(c6:g6*c7:g7)* and **<Ctrl> + <Shift> + <Enter>**

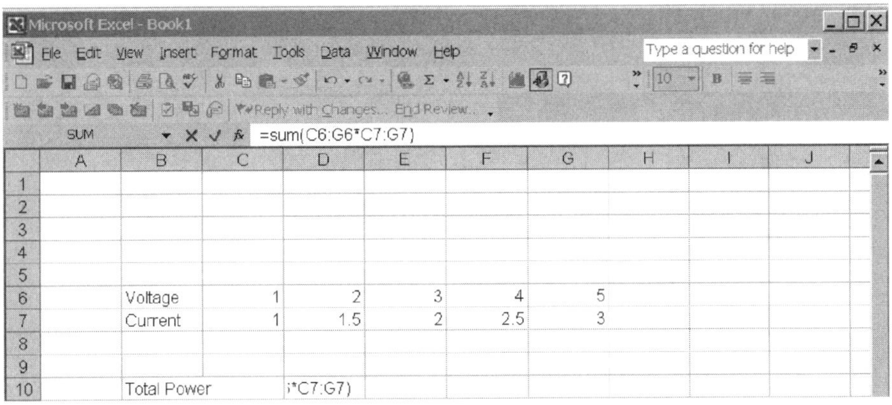

Figure A-8: Array formula with results stored in memory.

Returning an Array from a Function

A very powerful feature of VBA within Excel (and possibly unique to any programming language), is the fact that Excel is capable of returning an array of values from a function call and not just a single value. Remember, when entering the formula in the worksheet to use the <Ctrl> + <Shift> + <Enter>.

Error Values

When an Excel function detects an error, it returns a description that cryptically provides a diagnosis of the problem. Possible error descriptors are:

Error Descriptor	Diagnosis
#VALUE	One of the calling parameters to the function is incorrect.
#NULL	There is no intersection (of course, Excel must be looking for the intersection) of the selected ranges.
#REF	Invalid cell reference (cell being accessed has been deleted).
#NAME?	Cell/Range name reference error. It has been renamed or the name deleted.
#NUM	A value somewhere is not what was expected. Perhaps a number outside of the permissible range.
#NA	The function is trying to access a cell where the data is invalid or not available.
#DIV/0!	The path to "infinity and beyond" is blocked by this Error Message (with acknowledgement to Buzz Lightyear).

If you want your function to return an error value that will comply with Excel standards so that Excel functions act in a consistent way, you need to assign an error code to the return value using the CVErr function. The code may look something like this:

```
If  InputParameter<0 Then
      FunctionName=CVErr(xlErrNum)
Else
     ...
```

The possible constants corresponding to the Error Descriptor above are:

Error Descriptor	Error Constants
#VALUE	xlErrValue
#NULL	xlErrNull
#REF	xlErrRef
#NAME?	xlErrName
#NUM	xlErrNum
#NA	xlErrNA
#DIV/0!	xlErrDiv0

Useful Functions

Here are some useful features, just to give an indication of what is possible. Using the VBA Help (and the **See also** links) will provide more detailed usage.

Dir (filename) will search for a filename (including path), returning a null string if it does not exists.

GetAttr (PathName) will return an integer representing the attributes of a file, directory, or folder.

ActiveWorkbook.Names is a method of accessing the named ranges of the active workbook.

ActiveWorkbook.Sheets (SheetName) is a method of detecting if there is a sheet of a particular name in the active workbook. It returns an error if there is no sheet.

Workbooks returns all the workbooks that are open.

Workbooks("FredNurke").Activate will activate the named workbook.

Workbooks.Open filename:="RTD.XLS" will open the named workbook.

The *cell* function is actually an Excel function, but I am not quite sure where to include it. It packs quite a punch, enabling you to find out everything about a cell and more, including the path of the workbook that it is found in. Have a look at the Excel Help on the subject.

Web Resources

Aside from the Microsoft Knowledge Base and the information in *PC Magazine*, there are three "excel"lent web sites dedicated to Excel that I know about. Although they are more oriented towards the mass market, there are many gems in these sites. Enjoy your digging.

1. http://j-walk.com/ss/excel/index.htm
2. http://www.mrexcel.com/articles.shtml
3. http://www.cpearson.com/excel.htm

Parallel and Serial I/O

Based on the number of packages I found that allow Excel to interface to the real world (Table 1), there is a market for this type of application. Indeed, I have needed to do this several times, but every time I tried it, I never really found a way to do it without having to spend several hundred dollars. I always felt there was a way, and in researching this book, I think I can show you how to interface to the parallel and serial ports without too much expense. Unfortunately, this information comes right at the end of the life of the parallel and serial ports. For some time into the future, however, engineering and scientific instrumentation are likely to have serial ports, so if your PC does not have a serial port you should be able to interface to a serial port with a USB-to-serial port adapter. The parallel port is handy for some low-level electronic prototyping or creating some kind of electronic circuitry that occasionally appears in the design ideas of electronics magazines. Most of these ideas use C to achieve the software interface between PC and prototype. The techniques used here will work in VB as well, but because this book is about Excel, we will be using VBA.

Table B-1 details all the information I could find on the Internet about interfacing to parallel and serial ports. Some of the information is dated, some may no longer be there by the time this book is published, some of them may or may not work on your machine. No doubt, I have missed many. I played with most of the freeware and some of the shareware. I also tried Measure® from National Instruments.

There appear to be several ways of creating the software interface within VBA. It is not always obvious to me what technique is being used in the applications listed in Table B-1. I think all of the possible ways are as follows:

DLL: This is simply a module that you treat as external calls. The DLL file must appear in the system or system32 subfolder of the windows folder.

Sendkeys: It is possible to send a message between applications as if the keyboard was being pressed using the Sendkeys function. See the help in a Microsoft application for this approach.

Product Name	Manufacturer	Web site	Product type	Price
Direct-IO	Ingenieurbuero Paule	www.direct-io.com	DLL	$29 shareware
Comm-Drv/Lib	RegNow	http://development.newfreeware.com/programs/165/	DLL	$99.95
IO.dll	Geek Hideout	http://www.geekhideout.com/iodll.shtml	DLL	Free
CommX	Geenleaf Software	http://www.greenleafsoftware.com/Com-mX/CommXInfo1.asp	DLL	$249
CommTools/ DLL	Magna Carta Software	http://www.sofdesign.com/developer/commtools/dll.html	DLL	$249.95
Measure	National Instruments	www.ni.com	DLL	$495
WinWedge	Taltech	www.taltech.com	DDE/ SendKey	$259
Inpout32.dll	Logix4u	http://www.logix4u.cjb.net/	DLL	$240
CrystalCOMM	Crystal Software	http://www.crystalcom.com/crs_cnt.htm	DLL	$200
Windmill 4.3	Windmill Software	http://www.windmill.co.uk/rs232.html	?	Free
Port95nt	Scientific Software Tools	http://www.driverlinx.com/		
Measure Foundry	Data Translation	www.datatranslation.com	?	$995
COMxL RS232C	Lye Softlab	http://www.geocities.com/lyesoft/	?	$59
MSComm32	Microsoft	www.microsoft.com	DLL	Free?
UltimaSerial		http://www.geocities.com/ultimaserial/ May not be universal	DLL	$18 personal use

Table B-1. Software resources.

DDE: Microsoft created this standard to allow handshaking between two separate applications. It is my understanding that the approach is being replaced by RTD functionality.

RTD: (Real Time Data). This functionality is intended for the acquisition of data like stock prices, but it doesn't take much imagination to stretch this to any data. You just need an application to format the data. If you want to do this yourself, you will need to know about Component Object Model (COM) automation.

Windows Component Objects: It is possible for an application (written in Visual Basic, for example) to open Excel and modify components and objects as desired. You may want to do this if your application needs some specialized functionality like charting, but you need the flexibility of a separate application.

Support for lower-end software in Table B-1 is nonexistent. It is also not possible to predict whether something will work on some particular hardware since the hardware I/O map has changed from the original LPT and COM days. In addition, the newer operating systems like Windows 2000 try to isolate the application from direct I/O writing. You just have to fiddle around until you find something that works for you. The two solutions that I use in the book probably have a greater chance of succeeding, but I can't promise.

For parallel I/O, I am using IO.DLL available from Geek Hideout (http://www.geekhideout. com/iodll.shtml). The author, Fred Bulback, has graciously allowed me to supply the file on the CD. I claim no credit, nor take any responsibility for it. You can find an example using IO.DLL (besides mine of course) from http://www.southwest.com.au/~jfuller/vb/vbout.htm.

I did try to use IO.DLL (and many others) to directly drive the serial port, but to no avail. Perhaps this is just as well, since it would have been difficult to create an interrupt handler.

For the serial I/O I am going to use the MSComm ActiveX control that has been used in Visual Basic.

Parallel (Printer) Port

A good source for all things to do with the parallel port is the web site associated with the book, *Parallel Port Complete*, by Jan Axelson (www.lvr.com).

In order to install IO.DLL, all you have to do is locate the file on the web and download it. The DLL file should be placed in c:\windows\system32 or c:\windows\system or similar. The public declarations (copied directly from the web site) are described in the model in the main body of the book.

Serial Port

In order to use the MSComm ActiveX control, you need to have the file MSComm32x.ocx in your "system" or "system32" subfolder. However, this does not guarantee that Excel will recognize it. I think this file comes with the Professional and Enterprise versions of Microsoft Office, but I can't guarantee that. It also comes with Visual Basic in the Professional and Enterprise versions. However, you can get it on the web as well. Try www.zaber.com and http://www.yes-tele.com/mscomm.html. If it is not on your hard drive you must obtain it. It is not easy to tell what does work because once MSComm has been registered on your system, it can't be undone. So it was impossible for me to try something else to test different approaches.

In order to detect whether you have MSComm, first let's investigate how to place the communications control in VBA. Open Excel and go to VBA (<Alt> + <F11>). Then click on **Insert | User Form** and you should see Figure B-1.

Right click on the toolbox and select **Additional Controls** form the menu. Scroll down and find the control "Microsoft Communications Control" as in Figure B-2. Click on the selection box and on **OK**.

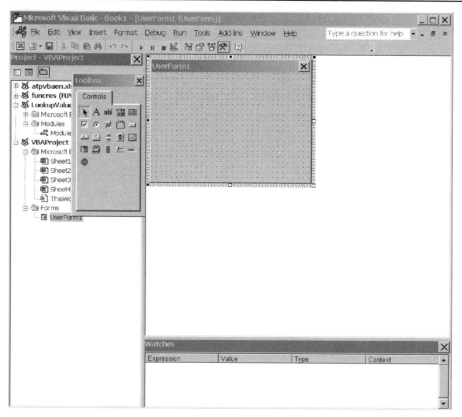

Figure B-1: Inserting a user form. If the Toolbox is not visible, enable it from the View menu.

A telephone icon will should appear on the toolbox. Click on it and then click on the user form and drag a rectangle. If you see the telephone on the form, then mutter a quiet prayer of thanks and ignore the next few paragraphs.

Figure B-2: Adding the serial port control.

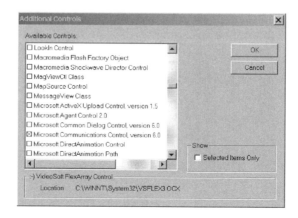

In all probability you will get a message that this control is not licensed. How to license it? Well if you have Visual Basic (probably VB6) then bypass the next few attempts.

First try to register the software. Click on the **Start** button, then **Run** and type in:

   regsvr32 mscomm32.ocx

Now try to place the communications control again. Successful? I didn't think so.

Try to download VBDEMO.ZIP from www.zaber.com. Install and run it as per Zaber's suggestion. Did it work for you? Me neither. (While you are here, download ZaberExcelVBdemo for some examples on MSComm in Excel.)

In Microsoft VBA, click on **Tools** | **References** and **Browse** and find the ActiveX Control mscomm32.ocx. Does it work now? Well, it was a long shot. I tried finding out from Microsoft, but they claim it is not an installation problem and that I should try paying to get advice. After my experience with the User function, where I was charged $35 to find out that nobody could give me the answers I needed, I decided to take the advice given at http://www.yes-tele.com/mscomm.html and I gave up. Incidentally, this site has the only full description of the MSComm control that I have found.

The only sure-fire way I have found is to install VisualBasic 6 Professional or Enterprise editions. I don't really know what will happen with earlier or later versions of Excel vis-à-vis the Visual Basic version since once the control is registered it stays that way, no matter what. You can uninstall it immediately afterwards, but the registration remains. If I were you I wouldn't try this without consulting a lawyer, but it seems to me that you could borrow VB6, install and then immediately uninstall it and you wouldn't be violating the copyright.

If you want to get VB6 Professional or better, the easiest (and cheapest) way that I found is to buy VB.NET. I can only find the standard edition listed anywhere. Microsoft will allow you to downgrade any software you buy, so once you have VB.NET, contact Microsoft and ask for the VB6 Professional or Enterprise CDs. Remember to ask for the MSDN library CDs as well if you intend to use VB6 at all.

Possible Hardware Sources

USB to serial and/or parallel ports can be found at SeaLevel Systems (http://www.sealevel.com) (Tech support there actually suggested MSComm as a potential solution), or B & B Electronics (http://www.bb-elec.com). National Instruments and Belkin also make adapters.

References

In as much as any author is a product of experience and absorbed knowledge, the following books and articles have contributed some part to the contents of this book. It is sometimes difficult to provide a one-to-one traceability, but I am grateful to all the authors who have shared their wisdom with the world. The books in **bold** have featured prominently in my learning of Excel.

1. *Do You Excel in Electronics?* Circuit Cellar Online January–March 2002.
2. *Microsoft Excel User's Guide Version 5.0*, Microsoft Corporation © 1992–1993.
3. *Microsoft Office 2000 Visual Basic Programmer's Guide*, Microsoft Corporation © 1991–1999.
4. **Using Microsoft Excel 97**, Bruce Hallberg, Sherry Kinkoph, Bill Ray, et al., Que Corporation © 1997, ISBN 0-7897-1399-3.
5. *Ease Excel Input*, Helen Bradley, PC Magazine, March 12, 2001.
6. *Create Your Own Excel Add-In*, Helen Bradley, *PC Magazine*, December 24, 2001.
7. *Microsoft Excel Tips*, PC Magazine, April 9, 2002.
8. *Excel Offers Painless LCD Initialization*, Alberto Ricci Bitti, Design Ideas, *EDN*, September 20, 2001.
9. *Tricks Improve on Excel LCD Initialization*, Aubrey Kagan, Design Ideas, *EDN*, April 11, 2002.
10. *Driving the NKK Smartswitch*, Aubrey Kagan, Circuit Cellar, July/August 2002.
11. *Serial Port Complete*, Jan Axelson, Lakeview Research, ©1998, ISBN 0-9650819-2-3.
12. *Visual Basic Programmer's Guide to Serial Communications*, Richard Grier, Mabry Software Inc., ©1997-8, ISBN 1-890422-25-8.
13. **Microsoft Excel 2000 Power Programming with VBA**, John Walkenbach, IDG Books Worldwide Inc., ©1999, ISBN 0-7645-3263-4.
14. **Excel Programming**, Jinjer Simon, Wiley Publishing, ©2002, ISBN 0-7645-3646-X.
15. **Writing Excel Macros with VBA**, Steven Roman, O'Reilly & Associates, Inc., ©2002, ISBN 0-596-00359-5.
16. **Excel Charts, John Walkenbach**, Wiley Publishing, Inc., ©2003, ISBN 0-7645-1764-3.
17. **Microsoft Excel: Tips, Techniques and Shortcuts**, Rockhurst University Continuing Education Center, Inc., ©2001.
18. **Advanced Microsoft Excel for the Power User**, Rockhurst University Continuing Education Center, Inc., ©2000.

19. *Quick Reference Software Guide: Visual Basic 6.0*, Bar Charts Inc., ©2000, ISBN 157222374-X.
20. *Using Visual Basic 6*, Brian Siler and Jeff Spotts, Que Corporation, ©1998, ISBN 0-7897-1542-2.
21. *The Beginner's Guide to Visual Basic 4.0*, Peter Wright, Wrox Press Ltd, ©1995, ISBN 1-874416-55-9.
22. *Handle File I/O*, Ron Schwarz, *Visual Basic Programmers Journal*, December 2000.
23. *Programmer's Guide to Microsoft Visual Basic, Programming System for Windows*, Version 4.0, Microsoft Corporation, ©1995.
24. *Word97 Macro and VBA Handbook*, Guy Hart-Davis, Sybex Inc., ©1997, ISBN 0-7821-1962-X.
25. *Visual Basic 5 Developer's Handbook*, Evangelos Petroutsos and Kevin Hough, Sybex Inc., ©1998, ISBN 0-7821-1985-9.
26. *Visual Basic 5: Object Oriented Programming*, Gene Swartzfager, The Coriolis Group, ©1997, ISBN 1-57610-106-1.
27. *Programming Microsoft Visual Basic 6.0*, Francesco Balena, Microsoft Press, ©1999, ISBN 0-7356-0558-0.
28. *Visual Basic 5 Superbible*, Eric Winemiller et al., The Waite Group, ©1997, ISBN 1-57169-102-2.
29. *Circuit Puts Analog Data into Excel*, Clayton B. Grantham, Test and Measurement World, September 1996.
30. *Add Voice Command to Virtual Instrumentation*, Alexander Bell, Design Idea, *EDN*, May 30, 2002.
31. *Add Voice Command to Your CAD System*, Alexander Bell, Design Idea, *EDN*, May 2, 2002.
32. *Design Low-Duty-Cycle Timer Circuits*, Phil Rogers, Design Idea, *EDN*, August 22, 2002.
33. *Voice Feedback Enhances Engineering Calculator*, Alexander Bell, Design Idea, *EDN*, July 11, 2002.
34. *Automate Immunity Tests with Excel*, Ken Hall, *Test and Measurement World*, November 1, 2002.
35. *Trace Voltage-Current Curves On Your PC*, Clayton Grantham, *Test and Measurement World*, October 11, 2001.
36. *Track Multisite Temperatures On Your PC*, Clayton Grantham, Design Idea, *EDN*, April 18, 2002.
37. *Excel 2002 For Windows*, Global Reference Guides Inc., ISBN 1-55353-028-4.
38. *Conditional Formatting in Excel*, Neil J. Rubenking, *PC Magazine*, April 8. 2003.
39. *Reliability Prediction Procedure for Electronic Equipment*, Bellcore Technical Reference TR-NWT-000332, September 1990.
40. *An Elementary Guide to Reliability*, Second Edition, G.W.A. Dummer and R.C. Winton, Pergammon Press, ISBN 0-08-017821-9, 1974.
41. *MCS-51 Macro Assembly Language Pocket Reference for DOS Systems*, Intel Corporation, 122755-001, 1986.
42. *Fine-Tune Your Office XP Settings*, www.zdnetindia.com/help/howto/stories/24577.html, Gregg Keizer, June 4, 2001.
43. *Customize Speech Recognition Voice Commands in Office XP*, Microsoft Office Assistance Center, http://office.microsoft.com/assistance/2002/articles/oWebCustomizeSpeechVoiceComm.aspx.
44. *XL2000: Macro to Import a Text File into an Existing Worksheet*, Microsoft Knowledge Base Article 213816, http://support.microsoft.com.

45. *Managing Macros with the Visual Basic Editor*, Paul Cornell, http://office.microsoft.com/assistance/2002/articles/pwUsingTheVBE.aspx.
46. *Leaded Fixed Linear Resistors*, 1998/9 Data Handbook PA08b, Philips Components.
47. *RTD tables: Instrumentation.com*, http://www.instrumentation.com.
48. *Practical Temperature Measurements*, Application Note 290, Hewlett-Packard.
49. *XTR105 Data Sheet*, SBOS061A- May 2003, Texas Instruments, www.ti.com.
50. *LM117/LM317A/LM317 3-Terminal Adjustable Regulator Data Sheet*, National Semiconductor, May 2003, www.national.com.
51. *Sil-Pad ® K-10 Data Sheet*, PDS-0602-001-01; Rev 01, The Bergquist Company, www.bergquistcompany.com.
52. *Electrical Engineering Science*, Preston R. Clement & Walter C. Johnson, McGraw-Hill Book Company, Inc. Library of Congress Catalog Card Number 59-15457, 1960.
53. *TL431, TL431A Adjustable Precision Shunt Regulators Data Sheet*, Texas Instruments, SLVS005U, July 1978, revised July 2003, www.ti.com.
54. *www.frontsys.com*, Frontline Systems, Inc. (Creator of Solver.)
55. *Quattro Pro 10 Help File*, Corel Corporation 2001, www.corel.com.
56. *The 555 Timer © 119*, 2000-2002 Ken Bigelow, http://www.play-hookey.com.
57. LM555 timer datasheet, February 2000, National Semiconductor Corporation, www.national.com.
58. *AN170 NE555 and NE556 Applications*, December 1998, Philips Semiconductors.
59. *ICM7555, ICM7556 Data Sheet*, November 2002, Intersil, www.intersil.com.
60. *ICM7555/7556 data Sheet*, November 1992, Maxim, www.maxim-ic.com.
TLC555 LinCMOS Timer, September 1983, revised March 2001, Texas Instruments, www.ti.com.
62. *MC1455, MC1455B, NCV1455B Timers Data Sheet*, January 2003, ON Semiconductor, http://onsemi.com.
63. *LM555/NE555/SA555 Single Timer*, 2002, Fairchild Semiconductor Coporation, www.fairchildsemi.com.
64. *Combo Boxes in Excel*, Neil J. Rubenking, PC magazine, Aug 19, 2003
65. *Implement a Nine-Data-Bit UART on a PC*, Aubrey Kagan, Design Idea, *EDN* June 4, 1998.
66. *Microsoft Visual Basic 4.0 Developer's Workshop*, John Clark Craig, Microsoft Press 1996, ISBN 1-55615-664-2
67. *Absolute and Relative References in Excel*, PC Magazine, October 14, 2003, Neil J. Rubenking.
68. *AD736 Data Sheet*, Rev E, Analog Devices.
69. *AD637 Data Sheet*, Rev E Analog Devices.
70. *RMS-to-DC Application Guide, Second Edition*, Analog Devices.
71. *Entering Nonstandard Characters*, PC Magazine, August 19, 2003, Neil J. Rubenking.

About the Author

Born and raised in what is now Zimbabwe, Aubrey Kagan gained an electrical engineering degree in Israel and an M.B.A. in South Africa. He has lived in Toronto, Canada for the past 15 years where he is a licensed professional engineer. His 25 years of experience in South Africa and Canada encompass the use of electronics in industrial and mining applications, although for a brief stint he worked on the requirements for the Canadian built robotic arm now installed on the International Space Station.

He has had four design ideas and six articles published in technical and professional magazines. Two of these were the seeds for this book.

Index

TERM

This Agreement will remain in effect until terminated pursuant to the terms of this Agreement. You may terminate this Agreement at any time by removing from Your system and destroying the CD-ROM Product. Unauthorized copying of the CD-ROM Product, including without limitation, the Proprietary Material and documentation, or otherwise failing to comply with the terms and conditions of this Agreement shall result in automatic termination of this license and will make available to Elsevier Science legal remedies. Upon termination of this Agreement, the license granted herein will terminate and You must immediately destroy the CD-ROM Product and accompanying documentation. All provisions relating to proprietary rights shall survive termination of this Agreement.

LIMITED WARRANTY AND LIMITATION OF LIABILITY

NEITHER ELSEVIER SCIENCE NOR ITS LICENSORS REPRESENT OR WARRANT THAT THE INFORMATION CONTAINED IN THE PROPRIETARY MATERIALS IS COMPLETE OR FREE FROM ERROR, AND NEITHER AS-SUMES, AND BOTH EXPRESSLY DISCLAIM, ANY LIABILITY TO ANY PERSON FOR ANY LOSS OR DAMAGE CAUSED BY ERRORS OR OMISSIONS IN THE PROPRIETARY MATERIAL, WHETHER SUCH ERRORS OR OMIS-SIONS RESULT FROM NEGLIGENCE, ACCIDENT, OR ANY OTHER CAUSE. IN ADDITION, NEITHER ELSEVIER SCIENCE NOR ITS LICENSORS MAKE ANY REPRESENTATIONS OR WARRANTIES, EITHER EXPRESS OR IMPLIED, REGARDING THE PERFORMANCE OF YOUR NETWORK OR COMPUTER SYSTEM WHEN USED IN CONJUNCTION WITH THE CD-ROM PRODUCT.

If this CD-ROM Product is defective, Elsevier Science will replace it at no charge if the defective CD-ROM Product is returned to Elsevier Science within sixty (60) days (or the greatest period allowable by applicable law) from the date of shipment.

Elsevier Science warrants that the software embodied in this CD-ROM Product will perform in substantial compliance with the documentation supplied in this CD-ROM Product. If You report significant defect in performance in writing to Elsevier Science, and Elsevier Science is not able to correct same within sixty (60) days after its receipt of Your notification, You may return this CD-ROM Product, including all copies and documentation, to Elsevier Science and Elsevier Science will refund Your money.

YOU UNDERSTAND THAT, EXCEPT FOR THE 60-DAY LIMITED WARRANTY RECITED ABOVE, ELSEVIER SCIENCE, ITS AFFILIATES, LICENSORS, SUPPLIERS AND AGENTS, MAKE NO WARRANTIES, EXPRESSED OR IMPLIED, WITH RESPECT TO THE CD-ROM PRODUCT, INCLUDING, WITHOUT LIMITATION THE PROPRIETARY MATERIAL, AN SPECIFICALLY DISCLAIM ANY WARRANTY OF MERCHANTABILITY OR FITNESS FOR A PARTICULAR PURPOSE.

If the information provided on this CD-ROM contains medical or health sciences information, it is intended for professional use within the medical field. Information about medical treatment or drug dosages is intended strictly for professional use, and because of rapid advances in the medical sciences, independent verification f diagnosis and drug dosages should be made.

IN NO EVENT WILL ELSEVIER SCIENCE, ITS AFFILIATES, LICENSORS, SUPPLIERS OR AGENTS, BE LIABLE TO YOU FOR ANY DAMAGES, INCLUDING, WITHOUT LIMITATION, ANY LOST PROFITS, LOST SAVINGS OR OTHER INCIDENTAL OR CONSEQUENTIAL DAMAGES, ARISING OUT OF YOUR USE OR INABILITY TO USE THE CD-ROM PRODUCT REGARDLESS OF WHETHER SUCH DAMAGES ARE FORESEEABLE OR WHETHER SUCH DAMAGES ARE DEEMED TO RESULT FROM THE FAILURE OR INADEQUACY OF ANY EXCLUSIVE OR OTHER REMEDY.

U.S. GOVERNMENT RESTRICTED RIGHTS

The CD-ROM Product and documentation are provided with restricted rights. Use, duplication or disclosure by the U.S. Government is subject to restrictions as set forth in subparagraphs (a) through (d) of the Commercial Computer Restricted Rights clause at FAR 52.22719 or in subparagraph (c)(1)(ii) of the Rights in Technical Data and Computer Software clause at DFARS 252.2277013, or at 252.2117015, as applicable. Contractor/Manufacturer is Elsevier Science Inc., 655 Avenue of the Americas, New York, NY 10010-5107 USA.

GOVERNING LAW

This Agreement shall be governed by the laws of the State of New York, USA. In any dispute arising out of this Agreement, you and Elsevier Science each consent to the exclusive personal jurisdiction and venue in the state and federal courts within New York County, New York, USA.